T0192298

Advanced Product Quality Planning

PRACTICAL QUALITY OF THE FUTURE

What it Takes to be Best in Class (BIC)

SERIES EDITOR

D. H. Stamatis

President of Contemporary Consultants
MI, USA

Quality Assurance
Applying Methodologies for Launching New Products, Services, and Customer Satisfaction
D. H. Stamatis

Advanced Product Quality Planning
The Road to Success
D. H. Stamatis

For more information about this series, please visit:
https://www.crcpress.com/Practical-Quality-of-the-Future/book-series/PRAQUALFUT

Advanced Product Quality Planning

The Road to Success

D. H. Stamatis

CRC Press is an imprint of the
Taylor & Francis Group, an **informa** business

CRC Press
Taylor & Francis Group
6000 Broken Sound Parkway NW, Suite 300
Boca Raton, FL 33487-2742

First issued in paperback 2020

ISBN-13: 978-1-138-39458-2 (hbk)
ISBN-13: 978-0-367-78070-8 (pbk)

Library of Congress Cataloging-in-Publication Data

Names: Stamatis, D. H., 1947- author.
Title: Advanced product quality planning : the road to success / D.H. Stamatis.
Description: Boca Raton, FL : CRC Press/Taylor & Francis Group, 2018. | Includes bibliographical references and index.
Identifiers: LCCN 2018034535| ISBN 9781138394582 (hardback : acid-free paper) | ISBN 9780429401077 (ebook)
Subjects: LCSH: Manufacturing processes--Quality control. | Product design.
Classification: LCC TS156 .S73445 2018 | DDC 670--dc23
LC record available at https://lccn.loc.gov/2018034535

Visit the Taylor & Francis Web site at
http://www.taylorandfrancis.com

and the CRC Press Web site at
http://www.crcpress.com

Contents

List of acronyms

1PP	first production prove-outs
AAR	appearance approval report
AIAG	auto industry action group
a.k.a	also known as
ANFIA	Associazione Nazionale Filiera Industria Automobilistica/Italy
APPC	average purchased part capacity
APQP	advanced product quality planning
APW	average production weekly
AQP	advanced quality planning
ARL	attribute requirements list
ARL	average run length
AWM	automotive warranty management
BIQS	built in quality supply base
CA	capacity analysis
CA	corrective action
CC	critical characteristic
CFT	cross-functional team
CIR	customer input requirements
CNs	change notices
COA	certificate of analysis
CP	control plan
CPCP	chrysler product creation process
C_{pk}	capability index of a stable process—short term
CPR	cost per repair
DFM/A	design for manufacturing and assembly
DFR	decreasing failure rate—a downward bending curve on an AWS hazard plot for TIS or mileage
DRIVe	delivery rating improvement verification
DV&R	design verification and reporting
DVP&R	design verification plan and report
DVP	design verification plan

E-108	branding directive. Ford global automotive parts trademarks
EBSC	external balanced scorecard
ELV	end of life of vehicle
EPV	error proofing verification
ES	engineering specifications
EWT	early warranty tracking
FECDS	Ford engineering CAD and drafting standards
FEU	field evaluation units—a fleet of saleable evaluation units built on the assembly line at 12 weeks before Job #1
FIEV	Fédération des Industries des Équipements pour Véhicules/France
FMEA	failure mode and effect analysis
FPDS	Ford product development system
FPSC	first production shipment certificate
FSP	Ford supplier portal
FTA	fault tree analysis
FTT	first time through
Gauge R&R	gauge repeatability and reproducibility
GCS	global claims system
GD&T	geometric dimensioning and tolerancing
GIM	global issue management
GME	general motors Europe
GPDS	global product development system
GYR	green, yellow, red
HTIS	part per vehicle issues—high time in service
IAOB	International Automotive Oversight Bureau/US
IATF	International Automotive Task Force/France
IFR	increasing failure rate—an upward bending curve on a AWS hazard plot for TIS or mileage
IPD	in plant date
IRE	initial risk evaluation
KIP	key input processes
KO	kick off date (for a particular program)
KPI	key process indicators
KPI	key process input
KPO	key process output
LPA	layered process audit
MAQMSR	automotive quality management system document
MAQMSR	minimum automotive quality management system requirements
ME	manufacturing engineering
MMOG/LE	material management operations guideline/logistics evaluation

MP&L	material planning and logistics
MPPC	maximum purchased part capacity
MPW	maximum production weekly
MR	model responsible
MRD	material review date
MSA	manufacturing site assessment
MSA	measurement system analysis
NBH	new business hold
NCT	non-conformance tracking
NIST	National Institute of Standards and Technology
NTEIs	new tooled end items
OEE	overall equipment effectiveness
OEM	original equipment manufacturer
ASQ	American society of quality
PA	process audit
PA	program approval (For a particular program. Usually the funds for the program are approved)
PBCP	process-based business collaboration platform
PBCP	prototype build control plan
PCA	permanent corrective action
PCA	process control audits
PCP	product creation process
PD	product development
PDC	product design complete
PDCA	plan, do, check, act (known as the Shewhart cycle)
PDR	production demonstration run
PDSA	plan, do, study, act (known as the Deming cycle)
PFC	process flow chart
PND	program need date
PPAP	production part approval process
PPC	purchased part capacity
P_{pk}	performance index—long term. (P_{pk} is the preferred index for it accounts for the true standard deviation as opposed to the C_{pk} which uses the approximate value of (R-bar/d_2) can only be used for stable and normally distributed processes)!
PPQI	prototype process potential and quality indexes
PQOS	plant quality operating system
PPR	process planning review
PRAS	parts return analysis system
PSO	process sign-off
PSW	part submission warrant
PTC	production tooling complete
PV	process validation

PVP	process validation plan
PVT	plant vehicle team
QFD	quality function deployment
QMS	quality management system
QNA	quality narrative analyzer
QNBH	quality new business hold
RAT	recycling action team—a cross functional team working to increase the use of recycled material
RFQ	request for quote
RSMS	restricted substance management standard
SAWRP	supplier associated warranty reduction program
SC	significant characteristic
SC	strategic confirmation (for a particular program)
SCA	supplier change request
SCCAF	special characteristic communication approval form (A Ford Motor Company document that summarizes all critical and significant characteristics)
SI	strategic intent (for a particular program)
SIM	supplier improvement matrix
SMMT	Society of Motor Manufacturers and Traders/UK
SPC	statistical process control
SPDP	supplier preliminary data profile
SQE	supplier quality engineer
SQRA	supplier quality risk assessment
SRE	supplier readiness evaluation
SREA	supplier request engineering approval
STA	supplier technical assistant (Ford customer representative)
TACT	estimated cycle time
TAKT	actual cycle time
TAP issues	tooling aid process
TCD	target completion date
TGR	things gone right
TGW	things gone wrong
TKO	tooling kick-off
TPSL	top problem supplier location
VDA 6.3	a German approach that defines "a process-based audit standard" for evaluating and improving controls in a manufacturing organization's processes.
VDA	Verband Der Automobilindustrie E.V. (German Automobile Industry Association)
VDA-QMC	Verband der Automobilindustrie—Qualitäts Management Center/Germany

VFG	vehicle function group—code for component classification used in quality testing
VO	vehicle operations
VOC	voice of the customer
webCN	change notice system
WERS	worldwide engineering release ystem (Ford System)
WIS	warranty information system

Preface

Advanced product quality planning is a process developed in the late 1980s by a commission of experts gathered around the "Big Three" US automobile industry: Ford, GM, and Chrysler. Representatives from the three automotive original equipment manufacturers (OEMs) and the Automotive Division of American Society for Quality Control (ASQC, now called ASQ for American Society for Quality) created the Supplier Quality Requirement Task Force to develop a common understanding on topics of mutual interest within the automotive industry.

This commission invested five years in analyzing the then-current automotive development and production status in the US, Europe, and especially in Japan. At the time, the success of the Japanese automotive companies was starting to be remarkable in the US market.

APQP is utilized today by these three companies and some affiliates. Tier 1 suppliers are typically required to follow APQP procedures and techniques and are also typically required to be audited and registered to IATF 16949 by September of 2018. The success of this methodology has been met with enthusiasm in other industries and in European markets with appropriate and applicable adjustments to reflect the specific industries and requirements for the product and industry.

As important as APQP is, there are unfortunately no full-length books about it, perhaps because each OEM has so many individual requirements other than the AIAG's requirements. To be sure, there are many APQP training manuals even today, but hardly any books. In 1998, I published the first book on APQP with an intent to introduce the reader to the methodology and the management techniques necessary to fulfill the quality planning strategy. In this book, I continue the discussion of the core methodology but I also include detailed discussions about the key tools, a lengthy discussion on PPAP and an overview of the specific requirement of FCA, Ford and GM. Furthermore, I have addressed the requirements of the AIAG, VDA, IATF 16949 and ISO 9001 as they relate to APQP i.e. risk, warranty and GD&T.

The basis for the make-up of a process control plan is included in the APQP manual (AIAG: APQP). The APQP process is defined in the AIAG's

APQP manual, which is part of a series of interrelated documents that the AIAG controls and publishes. These manuals include:

- The failure mode and effects analysis (AIAG: FMEA) manual
- The statistical process control (AIAG: SPC) manual
- The measurement systems analysis (AIAG: MSA) manual
- The production part approval process (AIAG: PPAP 4th ed.) manual

The Automotive Industry Action Group (AIAG) is a non-profit association of automotive companies founded in 1982.

In our modern world we demand quality in everything that we do or receive. However, in demanding quality we are faced with a serious problem, as we do not have a standardized definition of quality. Generally, most of us will settle with a simple definition that states: *quality is defined by the customer.*

As simple as this definition is, it presents major problems because it is not standardized, and each customer may indeed have a different perspective of what quality is. To neutralize this vagueness most industries have come to recognize that fundamentally a "quality product or service" needs at least five elements for success in satisfying the customer, and these are:

1. Leadership commitment and engagement
2. Workers that are involved with their representatives
3. Business ethics and legality
4. Use of a systematic, comprehensive process to ensure effectiveness and continual improvement
5. Sustainability and integration

These five elements, if applied correctly and on a timely basis, will indeed provide some stability, accountability, sustainability, low risk of dissatisfaction, and satisfaction for a good product or service produced.

It turns out that these elements provide the basis for the Advance Product Quality Planning (APQP) process. To be sure, the process is pretty much standardized through the AIAG requirements, but there are still areas where specific organizations have their own.

The typical process of APQP is defined in five areas and each area has inputs and outputs. In a cursory form they are:

- Establishing the project
- Identifying legal and other requirements
- Defining scope, management commitments, and responsibilities
- Conducting internal audits against the predetermined milestones
- Certification procedure

In the automotive industry, the APQP is meant to cover all automotive OEM requirements for planning activities into one process. Suppliers apply APQP to ensure the quality of their new products and to drive continual improvement. It also provides a framework for a structured approach to product and process design. It represents a standardized set of quality requirements that enable suppliers to design a product that satisfies the customer. The primary goal of product quality planning is to facilitate communication and collaboration between engineering activities. As such, it requires the engagement of a cross-functional team (CFT) that includes marketing, product design, procurement, manufacturing, and distribution. The objective is to ensure a clear understanding of the voice of the customer (VOC), and to translate it into requirements, technical specifications, and special characteristics.

Obviously, one can see that in the final analysis, APQP provides a standardized way of sharing results between suppliers and customers (including, automotive companies), as well as guidelines for an effective development process, which for most organizations are the following "core" phases: development, industrialization, and product launch.

So, APQP's main content is to serve as a guide in the development process and also a standard way to share results between suppliers and customers (whoever they are). In these three core phases there are imbedded 23 generic topics that are monitored continually. These 23 topics are expected to be entirely completed before production is started. They cover such aspects as: design robustness, design testing and specification compliance, production process design, quality inspection standards, process capability, production capacity, product packaging, product testing and operator training plan, among other items. (It is important to realize that some organizations have more than 23. For example, Ford Motor Co. has 30 elements, 23 of which deal with quality and the last 7 deal with capacity and capability). A generic cursory overview of APQP shows that APQP focuses on:

- Up-front quality planning
- Determining if customers are satisfied by evaluating the output and supporting continual improvement

APQP consists of five phases:

- Plan and Define Program
- Product Design and Development Verification
- Process Design and Development Verification
- Product and Process Validation and Production Feedback
- Launch, Assessment and Corrective Action

These phases for practical purposes are translated into five major activities:

1. Planning
2. Product Development
3. Process Development
4. Product and Process Validation
5. Production

Finally, these major activities are subdivided into manageable elements. So, in the big picture of the APQP process we evaluate at least seven major elements. They are:

1. Understanding the needs of the customer
2. Proactive feedback and corrective action
3. Designing within the process capabilities
4. Analyzing and mitigating failure modes
5. Verification and validation
6. Design reviews
7. Control special/critical characteristics

Author

D. H. Stamatis is the president of Contemporary Consultants Co., in Southgate, Michigan. He is a specialist in management consulting, organizational development, and quality science. He has taught project management, operations management, logistics, mathematical modeling, economics, management and statistics for both graduate and undergraduate levels at Central Michigan University, University of Michigan, ANHUI University (Bengbu, China), University of Phoenix, and Florida Institute of Technology.

With more than 30 years of experience in management, quality training, and consulting, Dr. Stamatis has serviced numerous private sector industries including but not limited to: steel, automotive, general manufacturing, tooling, electronics, plastics, food, navy, department of defense, pharmaceutical, chemical, printing, healthcare and medical device.

He is a certified quality engineer through the American Society of Quality Control, certified manufacturing engineer through the Society of Manufacturing Engineers, certified Master Black Belt through IABLS, Inc. and he is a graduate of BSi's ISO 900 Lead Assessor training program.

Dr. Stamatis has written more than 70 articles, presented many speeches, and has participated in both national and international conferences on quality. He is a contributing author on several books and the sole author of 59 books.

Introduction

Sometimes there is confusion about the terminology used for APQP. The terms advanced quality planning (AQP) and advanced product quality planning (APQP) are used. So, let us examine each.

AQP is the generic methodology that focuses on the design/development of the supplier's manufacturing process to ensure that it is capable of producing parts that meet design requirements at the quoted tooling capacity. APQP, meanwhile, is a structured method for *defining* and *executing* the actions necessary to ensure that a product satisfies the customer. To be successful in AQP and APQP a team approach is necessary and open communication is mandatory—see Figure I.1.

Therefore, the goal of the APQP process is to facilitate communication between all persons and activities involved in a program and ensure that all required steps are completed on time, with a high quality-of-event, at

Figure I.1 The integration of communication and team effort in the APQP process.

acceptable cost and quality levels. In addition, the following characteristics must be accounted for:

- *Time*: Milestones for deliverables must be appropriate, applicable and doable.
- *Transparency*: No hidden agendas and cutting corners at the expense of quality initiatives.
- *Thoughtfulness*: Appropriate, applicable and a holistic approach must be given to particular situations within the set milestones for deliverables.
- *Tolerance*: Customers and suppliers must understand the value of correct tolerances and variability. Both must use appropriate and applicable measurement system analysis, to identify, correct and validate all measurements.
- *Capability and capacity*: Both must be understood by the customer and supplier. If not, both will have consequences down the road of delivery. Remember that capability identifies whether or not the *process* can produce what the customer wants. Capacity on the other hand, identifies whether or not the *product* can be delivered at the amount (quantity) level required within all specifications.
- *Trust*: Customers and suppliers must operate at all times with an attitude of win–win. If not, they will fall short of their quality goals. Trust is the result of determination to do an excellent job on everything that is required to complete the milestone. With this determination, it is also required to have the appropriate discipline and dedication for the specific completion of tasks required to complete the milestone. Finally, with trust, the responsibility of decisiveness is also of paramount importance as decisions must be appropriate, applicable to the task, and made within the confines of the set timing.

Special note: Although the terms AQP and APQP are very similar (with minor differences), in practice, they are used interchangeably.

So, the purpose of the APQP process is to establish:

- Common expectations for internal and external suppliers. Typical items of concern are:
  - Minimize/reduce late changes to the part and process
    - Reduce/eliminate quality spills at all stages of production
    - Reduce/eliminate warranty
    - Increase customer satisfaction
    - *In short, minimize risk, eliminate waste, and save money!*
- Common process metrics. These aid the facilitation and early identification of required changes. They also help uncover hidden issues

that may develop into potential problems. In essence, they help avoid late changes.

- A common (standardized) program status reporting format. Standardizing the reporting format adds a discipline to the process of APQP that facilitates the concept and practicality of continual improvement. It does that by focusing on: (a) providing a quality product on time, at acceptable cost, to satisfy customers; (b) providing definite roles and responsibilities for the APQP elements; and (c) a better understanding of how the APQP elements relate to organization's timing.

So, what do we need before we can begin AQP? The answer is very simple—but very difficult to implement consistently in any given organization due to constant changes within the organization and or outside the organization (changes in requirements of timing, specifications, material, and change in process are typical). In any case, the themes that define a good beginning of any APQP program are: (a) teamwork (must be cross-functional and multidisciplinary); (b) communication (must always be open and two ways); (c) timing and planning (must be appropriate, applicable and realistic); and (d) all activities must be identified—no hidden agendas.

The roles may vary depending on the customer, supplier, and or product. Typical roles may be from the following list: (a) program management, (b) quality and reliability, (c) team leaders, (d) engineers, (e) suppliers, and (f) program team. Specific content may be in the areas of: (a) sourced part or module (supplier must be known); (b) source package; (c) official program timing; (d) design model (as appropriate); (e) engineering specifications; (f) engineering change notices (CNs), if applicable; and (g) lessons learned from previous programs.

Benefits of APQP

APQP supports the never-ending pursuit of continual improvement. The first three sections of APQP focus on planning and prevention and make up 80% of the APQP process. The fourth and fifth sections support the remaining 20% of APQP and focus on validation and evidence. The fifth section specifically allows an organization to communicate its findings and provide feedback to further develop the standard and processes. A list of some of the benefits of APQP include:

- Direct resources by defining the vital few items from the trivial many
- Promote early identification of change
- Enable intentional change (what is being changed on purpose to bring value to the customer)

- Evaluate incidental change (in the environment, customer usage, degradation, and interfaces)
- Avoid late changes (post-release) by anticipating failure and preventing it
- Reduce design and process changes later in the product development process
- Produce on-time, quality product at the lowest cost
- Provide multiple options for mitigating the risk when found earlier
- Enable greater capability of verification and validation of a change
- Improve collaboration between design of the product and processes
- Improve the design for manufacturing and assembly (DFM/A)
- Provide selection of lower-cost solutions earlier in the process
- Enable legacy capture and reuse, advancement of tribal knowledge, and standard work creation and utilization
- Include best-practice APQP evaluation process, process metrics, and status reporting
- Link the APQP process to product development and manufacturing processes
- Define roles and responsibilities for the APQP process
- Develop a common APQP process for both internal and external manufacturing and assembly suppliers

Besides the requirements of ISO 9001 and IATF 16949, all these benefits make APQP appealing to organizations that are not part of the automotive industry but want to improve their design and development process (for more detailed information see Stamatis, 1998; Advisera, 2017).

How do we do it? A typical model for implementing APQP in any organization is shown in Figure I.2.

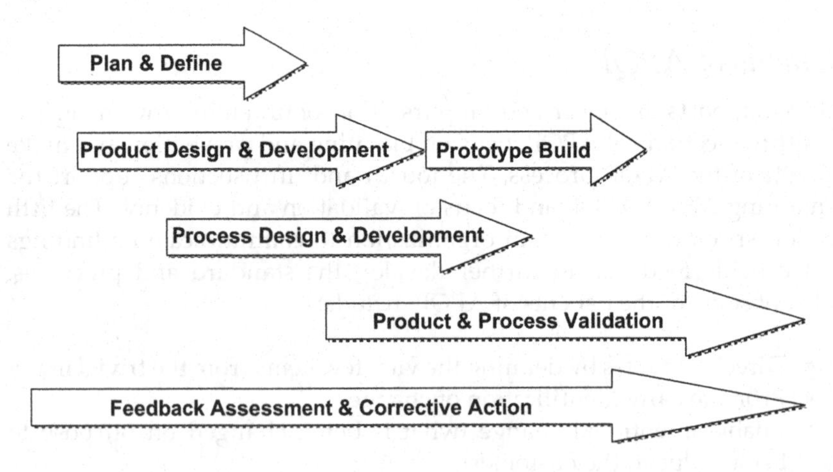

Figure I.2 A typical model for implementing APQP.

Table I.1 Key performance indicators

Concept Phase	Definition Phase	Target Setting Phase	Technical Design and Development Phase	Tooling Phase	Product Validation Phase
Start concept definition	Approve concept direction	Feasibility briefing/ pre spending approval; theme selection	Program and style approval; style freeze and tooling kickoff	Production design complete	Verification of process build

SOURCING (Purchasing Lead) ⟶
PRODUCT DESIGN (Engineering Lead) ⟶

The essence of the APQP process depends on certain foundations for the creation, identification and execution of all important milestones of key process variables (KPV). A typical definition of preset KPV milestones and timing may be in the form of Table I.1.

Key process indicators (KPIs) are standard quantifiable measurements that reflect the critical success factors of a program and designed so that they are measured consistently from program to program (repeatability). They are monitored throughout the development process within the program. It is very important to recognize that all KPIs are specific to each of the following timeframes:

- Approve concept direction to product design complete (PDC)
- Post-PDC to verification of process build
- Launch execution

To make sure that all KPIs are accounted for, a master list of KPIs should be generated with responsible areas identified appropriately. A typical master list may look something like Table I.2.

It must be noted here that in order to have successful KPIs a specific individual (or a team) must be responsible for each one. For example: (a) a designated lead is responsible for the execution of each KPI, (b) model responsible (MR) synthesizes all program KPIs, and (c) A designated product creation process (PCP) must be defined. The idea for this responsible party is to make sure that the defined process with its appropriate and applicable milestones will ensure that all requirements are delivered to the customer on time.

So, when one discusses APQP the fundamental essence is to make sure that the supplier's manufacturing is able to ensure capability of the producing parts. Of course, the ultimate insurance is depended on (a) process

Table I.2 A typical KPI master list

Program Plan		KPI Summary
KPIs	Timing of KPIs Based on process milestone (Gant Chart)	

Milestones are key decision points to validate program execution. Generally, there are two levels of milestones: (1) Major milestones—Area Level Cross-Functional Checkpoints (2) Secondary milestones—Area level Cross-Functional checkpoints (Exception: Theme select both senior and area level checkpoint).

sign-off (PSO) and or process validation (PV) testing, (b) Completion of PPAP and issuance of product submission warrant (PSW), and (c) PVP Build. Therefore, the results of an excellent APQP are:

- Minimize, reduce, or eliminate late changes by including the best practices of the APQP evaluation process, process metrics, and status reporting of past experiences.
- Reduce and/or eliminate quality issues at all stages of production by linking the APQP process to product development and manufacturing processes and by making sure that the milestones are doable and can be accomplished on time.
- Reduce and/or eliminate risk and warranty by defining the roles and responsibilities for the APQP process, especially in the areas of FMEA, PPAP, and prevention methodologies.
- Increase customer satisfaction by developing a common (standardized) APQP process for both internal and external manufacturing and assembly suppliers.
- In short, eliminate or reduce waste (however defined).

Obviously, the APQP not only depends on the specific program for specific milestones (gates of progress), but also on the scalability of the program. For example in the automotive industry that includes: (a) vehicle, (b) manufacturing, and (c) power train scalability. The reason why scalability is important is because there is a profound need of program review for the process used to monitor the status of the deliverables and the overall health of the program. This is accomplished by (a) uniform (standardized) reporting formats to communicate program status within the team structure and (b) two-way communications on all levels. Specifically, each of the five categories mentioned have generic inputs and outputs.

How does it work?

The aim of APQP is to enable an organization to achieve one task, and that is to develop a product quality plan for developing and producing products aligned with customer requirements. According to APQP standards, this planning uses a five-phase process. In a cursory format they are:

1. **Product planning and quality program definition:** In cases when the customer requires the introduction of a new product, or changes to an existing one, preliminary planning is the first step, far before discussion of the product design or redesign. The aim of this phase is to clearly define customer requirements and expectations regarding the product to be designed or redesigned. This phase includes obtaining necessary information and data to determine customer requirements. Once the requirements are identified, the quality program can be defined. The result of this phase is product design, reliability, and quality goals. Specifically, the inputs and output of this stage are:

   Inputs into Stage 1:
   - Voice of the customer
   - Market research
   - Historical issues
   - Team experience
   - Business plan and marketing plan
   - Product and process benchmark
   - Product and process assumptions
   - Product reliability studies
   - Customer inputs as applicable
   - Quality function deployment (QFD)

   Outputs of Stage 1:
   - Design goals
   - Reliability and quality goals
   - Preliminary bill of material (BOM)
   - Preliminary process flow
   - Preliminary list of special characteristics
   - Product assurance plan
   - Gateway approval

2. **Product design and development:** The goal of this phase is to finish the product design, as well as the feasibility assessment. Specifically, the inputs of this stage are the outputs of the first stage and its outputs are:

   Outputs of Section 2:
   - Design FMEA (DFMEA)
   - DFM/A
   - Design verification
   - Design review
   - Prototype control plan
   - Engineering drawings CAD the master
   - Engineering specifications
   - Material specifications
   - Change control for drawings
   - New equipment, tooling, and facilities requirements
   - Special product and process characteristics
   - Gauges/testing equipment requirements
   - Team feasibility commitment and gateway approval

3. **Process design and development:** This phase covers planning of the manufacturing process for the new or improved product. The goal is to consider product specifications, product quality, and production costs when designing and developing the production process. The process must be able to operate efficiently and produce the expected quantities to keep up with consumer demands. Specifically, the inputs of this stage are the outputs of the second stage and its outputs are:

   Outputs of Stage 3:
   - Packaging standards and specifications
   - Quality system review
   - Process flowchart
   - Floor plan layout
   - Characteristics matrix
   - Process FMEA (PFMEA)
   - Pre-launch control plan
   - Process instructions
   - Measurement systems analysis (MSA) plan
   - Preliminary process capability plan
   - Gateway approval

4. **Validation of product and process:** This is the test phase for validating the manufacturing process and the final product. The end results of this phase include confirming the capability and reliability

of the manufacturing process, and the product quality acceptance criteria. Trial production runs are made, and product output is tested to confirm the effectiveness of the deployed manufacturing approach. Any needed adjustments are reconciled before moving on to the next phase. Specifically, the inputs of this stage are the outputs of the third stage and its outputs are:

Outputs of Section 4:

- Significant production run
- MSA results
- Process capability studies
- Production part approval process (PPAP)
- Production validation testing
- Packaging evaluation
- Production control plan
- Quality planning sign-off and gateway approval

5. **Production launch, assessment, and improvement:** The full-scale production launch occurs in this phase, with an emphasis on evaluating and improving processes. Activities such as reducing process variations, identifying issues, and starting corrective actions to support continual improvement are mainstays in this phase. Collecting and assessing customer feedback, as well as collecting data related to process efficiency and quality planning effectiveness, is important in this phase. Specifically, the inputs of this stage are the output of the fourth stage and its outputs are:

Outputs of Section 5:

- Reduced variation
- Improved customer satisfaction
- Improved delivery performance
- Effective use of lessons learned

So, who is responsible for all these requirements? The responsibility for a complete APQP always lies with the supplier. However, in real terms the customer through their authorized representative such as the supplier technical assistance (STA) or the supplier quality engineer (SQE) as well as the supplier's team such as quality manager or quality engineer is involved throughout the entire process.

Risk assessment

As we already have mentioned, the APQP methodology is followed to prevent any problems in the fulfillment of the customer's requirements: it is a prevention methodology. As such, it is used as a thought starter to assist in assessing potential risk in a potential supplier. Therefore, quite often a risk assessment is necessary before (and sometimes during the execution of APQP) to help determine the extent to which the APQP status reporting process needs to be performed by the supplier—for a lengthy discussion on risk see Chapter 9. Risk assessment requires *subjective experience* to be used most effectively and that is why it is highly recommended that all risk assessments should have a checklist to follow especially if the participants are new to the methodology and or the product/process requirements. Typical situations where a risk analysis is pretty much required are the following:

- When a new supplier is being considered (sometimes new management will trigger an evaluation as well).
- When a new program in a current supplier is about to begin.
- When there is a new location or modified location.
- When a PSW is completed (at any time).
- When a new process or technology including new or substitute material is implemented.
- When a new product (commodity) or project is about to be commenced.
- When a current supplier has not shipped product for more than 24 months (the time here depends on the customer. The important thing is to recognize that some period has passed since the last delivery).

Check list guidelines

- Most planning for the APQP elements occurs from kickoff (KO) to strategic intent (SI) milestones.
- Program team reporting for most APQP elements begins at strategic confirmation (SC).
- Both the APQP checklist and status report have common fields that are *not* linked between the forms.
- The APQP checklist and status report must be updated for *each* milestone/team event as required by the program team or customer.

Reporting process

As with any program, so and with APQP there must be a process of reporting. A typical process may include the following steps:

1. *Planning of APQP elements.* Perhaps the most important stage of the entire process. Part of the planning is the kickoff meeting that will evaluate and define: (a) the initiation of the APQP process,

Table I.3 Risk/status assessment

Risk	Color	Definition
High	Red	Target dates and or deliverables are at risk and need immediate management attention. A recovery work plan is not available or implemented, or the work plan does not achieve program targets
Moderate	Yellow	Target dates and or deliverables are at risk, but are approved (team approved). Resource recovery work plan has been developed to achieve program targets and has been approved by the appropriate management team
None	Green	Target dates and deliverables are on track and the program element will meet the PND (program need date) and all quality objectives

(b) introduce the major concepts of the APQP, and (c) address the outcomes of the APQP.

2. *Execution of the APQP elements.* In this step the education (training) of the attendees takes place, as well as the specific roles and responsibilities of the major participants.
3. *Monitoring the "quality-of-event" of the execution of the elements, as well as the timing, by the team between the organizational milestones.* In this stage the monitoring of the task is followed very closely against the specific milestone and it is measured against the GYR (green, yellow, red) risk assessment metric—see Table I.3.
4. *Resolving issues.* In this stage we address any risk issues that may be encountered during the given milestone and try to resolve it either before the milestone deadline arrives or at least during the next milestone cycle arrives. If we miss the milestone the evaluation of red or yellow will still remain.
5. *Status reporting* (*minimum requirements are at organizational milestones*). In this stage we evaluate each task based on the completion stage that the specific milestone calls for with the GYR criteria. The criteria are shown in Table I.3.
6. *Discuss and finalize work plans.* Depending on the outcome of stages four and five, appropriate and applicable measures are taken to correct the issue(s) or problem(s).

The APQP status reporting procedural flow follows a generic path of five steps, which are:

Step 1: Complete APQP element for sphere of responsibility.
Step 2: Submit APQP element status to team leader.

Step 3: Team leader submits status to coordinator or quality manager.
Step 4: Coordinator or quality manager submits to SQE or STA.
Step 5: SQE or STA submits status to program SQE or program STA. The program SQE or STA summarize the data and process the information to customer's management team.

section one

The APQP process

chapter one

Plan and define

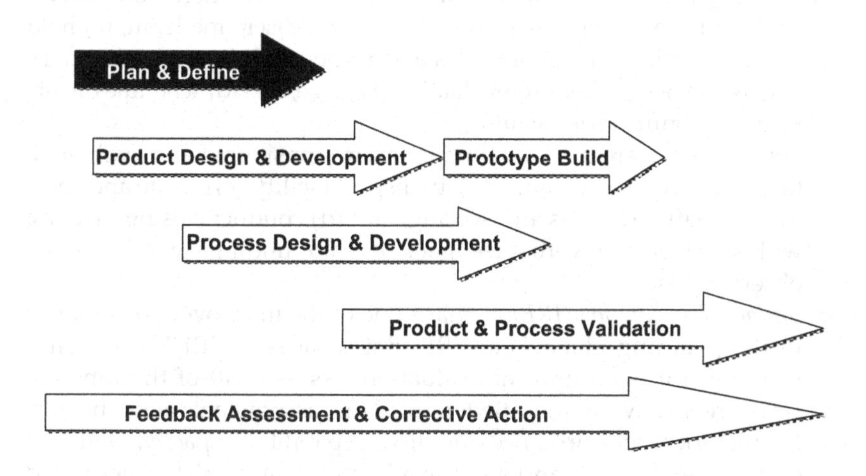

Perhaps one of the most important responsibilities of the entire advanced product quality planning (APQP) process is the plan and define stage. The essential responsibilities are:

Sourcing decision (SD)

In its most inherent responsibility it is "the" process that the customer uses to evaluate the supplier for future business. It serves as a formal customer commitment to work with all internal and external part suppliers, tooling suppliers, and facility suppliers on a program. Another way to describe it is that the SD is a coordinated effort between customers and suppliers for a "best fit" relationship to produce and accept the required product.

A critical evaluation of the potential supplier may be a source evaluation including a physical visit to the supplier's facility. A typical source evaluation has five components. They are:

1. *Identification*: Here the following items are considered and evaluated: (a) review list of possible suppliers; (b) analyze suppliers with existing ratings or evaluations—sometimes this analysis is based on surrogate data; (c) screen supplier pre-evaluation data; and (d) determine strategic suppliers needing evaluations.

2. *Task clarification*: Here the following items are considered and evaluated: (a) clarify scope of tasks related to product, site, and technology and supplier performance data, which may be from past experience with the supplier and or surrogate data from similar products; and (b) prepare decision-making recommendation and obtain management approval for the task.
3. *Visit preparation*: Here the following items are considered and evaluated: (a) form the cross-functional and multidiscipline team, (b) hold kickoff meeting and assign roles and responsibilities, (c) preliminary discussion of supplier pre-evaluation package, and (d) schedule on-site visit and confirm trip details.
4. *On-site visit*: Here the following items are considered and evaluated: (a) perform on-site evaluation, (b) inspect facility, (c) collate and coordinate analysis results and scoring, and (d) conduct closing meeting with supplier with verbal feedback of major findings (good and bad observations).
5. *Initial risk evaluation (IRE)*: Perhaps one of the most overlooked items in the planning phase is an IRE. But what is an IRE? In the simplest form it is an upfront evaluation (assessment) of the inherent risks in a new or modified part. This assessment may include: (a) historical performance (quality, capability, capacity, delivery, costs, safety, and manpower) availability; (b) available dedicated facilities; and (c) financial stability. (Here we must mention that the customer—depending on the status of the supplier—may require more than one IRE during the course of the program).

   So, what are some of the key risk drivers that should be assessed? Typical items of concern are summarized in Table 1.1.
6. *Results and reporting*: Here the following items are considered and evaluated: (a) finalize summary report and defined criteria, (b) calculate the evaluation's pass/fail rating and identify any risk factors, (c) report to approved managers, and (d) upload the information to the electronic system—if there is one.

At the end of this evaluation the key expectations are:

- All internal and external suppliers for a program are identified.
- Early sourcing target agreements are established and signed by appropriate areas.
- All nonspecific customer requirement (e.g., Q1 3rd ed.) suppliers are identified.
- Special arrangements for nonspecific customer requirement suppliers are identified.
- A program timeline, aligned to the organization's timing, is established from Strategic Intent to Job1.
- A risk assessment has been completed for all high-impact items.

Table 1.1 Typical items of concern in an IRE

Typical risk drivers	Y/N	Expectations	Guidelines
Are there issues preventing completion of tasks by required date?	A common acceptable scoring is a range of 60–80 and for safety items a range of 70–90. If color is used for evaluating risk the following is recommended: *RED*: Cannot bid/source business unless a customer director or higher signs off. *YELLOW*: Requires customer's senior manager signature for bid approval or sourcing business. *GREEN*: OK to bid or source business.	Review the planned completion dates for the tasks identified. Remember that the production tooling is complete by the verification of process time milestone.	Since tooling completion affects many APQP deliverables, it is imperative that it be monitored very closely. Late tooling should equate to a high completion after pre-volume production (PVP) and, if possible, it should be an automatic activity.
Will the production part approval process (PPAP) be completed by the required date?		Review the planned PPAP date. In the review you want to evaluate capacity projections.	If PPAP is not expected to be completed on time, a higher risk level is warranted. If engineering changes are the reason for the projected delay then the PPAP submission Delay (some call it: alarm; Request for Eng. Change and so on).

(Continued)

Table 1.1 (Continued) Typical items of concern in an IRE

Typical risk drivers	Y/N	Expectations	Guidelines
Is the design complete?		The part design/CAD should be complete and attached as part of the original source package.	If the supplier submitted a request for quote (RFQ) to an incomplete or non-existent part there is increased risk to the program. Tooling kickoff (TKO) without a complete design is also a high risk to the program.
Has historical design stability been considered?		Review lessons learned (both TGW and TGR) from previous or surrogate programs. Focus on the timing of engineering changes and whether they were minor or major in scope.	A high incidence of change notifications or one major or more than one minor change should reflect a higher risk level.
What is the customer impact if the part fails (in any way, or more specifically on issues of safety, application, functionality, etc.)?		Determine the potential customer impact if the part has failed in the past or it fails under the new conditions. Avoidance of any customer perception issues/problems must be a priority.	Parts with a high degree of customer interface should have been identified as high-risk items. Typical items here may be: safety components, appearance requirements for high visibility areas, etc.

(Continued)

Table 1.1 (Continued) Typical items of concern in an IRE

Typical risk drivers	Y/N	Expectations	Guidelines
What is the level of the process technology? Is it old? Widespread? Innovative?		Review the proposed manufacturing process. Make sure all functions are accounted for.	Processes that are using new technologies for the supplier require higher risk levels.
Is this a new manufacturing location (Greenfield)?		Manufacturing locations new to a customer must be evaluated and an acceptable score must be attained. All open issues—if there are any—must be closed. In case of any items being open, they must have appropriate overrides and authorization.	A new supplier to a particular customer represents a higher risk to the program since there is no direct history with that supplier, so the designated risk should be higher than for a supplier with some history (Brownfield).
Have there been historical quality problems with similar parts?		All historical quality characteristics in reference to a part—if applicable—or surrogate parts must be reviewed and evaluated to ensure these issues have been addressed so that they are not repeated.	Any high incidence for any key characteristic for a particular part represents a higher risk. Repeated issues or lack of corrective actions on any pending issue must be identified as a high-risk part or program.

(*Continued*)

Table 1.1 (Continued) Typical items of concern in an IRE

Typical risk drivers	Y/N	Expectations	Guidelines
What is the supplier's pre-assessment scoring method and what are the criteria for success, i.e., Q1 3rd ed. (Ford), Supplier Quality Bid List (SQBL; Fiat Chrysler)? Is there some form of specific score card for the supplier in addition to quality, cost, delivery, performance, risk? Or any other customer's system?			
Does the supplier have a warranty history, if yes, what is it? If not, why not?		Review the supplier's warranty history. Are they meeting their targets? Have they had any recent warranty or field actions?	If the supplier's warranty has been increasing, or if they have had any action recently, it may suggest additional risk precautions are needed.
Does the supplier have the ability to successfully conduct APQP activities?		Assess the supplier's ability to manage the APQP process and review the customer's key requirements. Evaluate and review the supplier location to determine whether or not multiple visits will be required based on past experience before receiving approval to proceed.	Make sure that sub-tier suppliers are guided and required to complete their own APQP and held responsible for their own PPAP.

Customer input requirements

The Customer Input Requirements (CIRs) element ensures that the program team appropriately assesses the potential manufacturing, process, and design issues early in the product development phase. In other words, the customer needs to know what they want. Once a program has commenced, any change is expensive and may challenge the preset timing of the program. Unquestionably, the CIRs are of profound importance since they are the design criteria and the program requirements necessary to initiate the APQP process. Key requirement expectations are:

- Issues from surrogate components or plants are identified and addressed with product and manufacturing.
- Component targets and product assumptions are defined.
- Targets for TGW, TGR, warranty, useful life reliability, and incoming quality are established.
- Capacity planning volumes are defined. Make sure that the overall equipment efficiency (OEE) is not reported more than 100%. If it is, there is a major error in the cycle time of the operation.
- System component designs and specifications should include:
  - Product assumptions
  - Functional performance
  - Dimensions
  - Weight
  - Material
  - Reliability and quality goals

Craftsmanship

Craftsmanship at this stage is considered a perception issue. However, even at this early stage it is a subjective assessment of what the customer sees, touches, uses, hears, and smells. It affects design and manufacturing, and it improves the overall perception of value. Key expectations are:

- Obtain craftsmanship targets and objectives to develop craftsmanship plans.
- Develop the craftsmanship strategy.
- Implement a plan to achieve craftsmanship targets.
- Integrate plans to support the final craftsmanship theme.
- Ensure that the design and manufacturing processes comply with craftsmanship strategy and plans.
- Identify and resolve craftsmanship issues that arise.

Team feasibility commitment

The team feasibility commitment element determines whether the proposed design can be manufactured within the program's guidelines and specifications. The goals of this element are to verify that all of the activities agree that they can produce the product within specifications. It is important to remember here that all feasibility concerns are resolved and necessary product and process changes are completed before first production prove-outs (1PP). Key commitment expectations are:

- The program team must establish a formal feasibility process and document.
- Feasibilities are tracked and major feasibility concerns can be resolved before the production trial run.
- Any concerns are resolved, and the product can be manufactured to meet Pp/Ppk and tolerance requirements.
- Suppliers and sub-suppliers were part of the feasibility process.

Note: Some alternative methodologies to planning are shown in Table 1.2.

Table 1.2 Some alternatives to APQP of planning

Juran's road map	Imai's plan		Nuclear industry's plan	Generic plan	Crosby's plan
• Identify customers • Discover customer's specific needs (concurrently, optimize product design) • Translate customer's needs into specifications (in your jargon) (concurrently, develop work process) • Establish measurement criteria (concurrently, prove process capability)	• Product planning • Product design • Product preparation • Purchasing • Production • Inspection • Sales and service	• Sales and engineering • Product planning and engineering • Product engineering, quality assurance, inspection, and production • Purchasing, inspection, quality assurance, and plant manager • Production management and plant manager	• Design products and parts • Procure control parts • Control materials brought into the work flow • Measure work processes • Calibrate measurement and test equipment • Develop, use, and control procedures for work and maintenance • Train workforce and manage knowledge for work and maintenance • Maintain all records related to quality	• Understand how the organization views quality • Understand customer's strategic needs and *desires* • Determine how quality is vital to the organization's mission • Align quality with organization's vision and values • Recognize how quality can help the entire organization in improvement endeavors	1. Establish clear policies about quality 2. Organize a cross-functional quality steering team 3. Train employees on quality methods 4. Establish measures and processes 5. Determine the price of non-conformance to specifications 6. Develop methods to share information about quality with the work force

(Continued)

Table 1.2 (Continued) Some alternatives to APQP of planning

Juran's road map	Imai's plan	Nuclear industry's plan	Generic plan	Crosby's plan
• Develop the product or service (concurrently, transfer knowledge as appropriate)	• Quality assurance and inspection • Sales management and service	• Develop and implement an audit process • Review and modify plan as needed	• Commit to using effective tools and methods in all aspects of the organization • Embrace a framework or management system (i.e., quality management system [QMS]) regarding strategic and tactical quality	7. Establish a systematic method for taking corrective action 8. Organize specific campaigns to achieve zero defects 9. Establish goals for improvement 10. Happy Zero Defects Day! 11. Enable employees to communicate to management about the causes of poor quality 12. Provide a recognition program for employees who improve quality 13. Establish local quality councils 14. Repeat

chapter two

Product design and development/prototype build

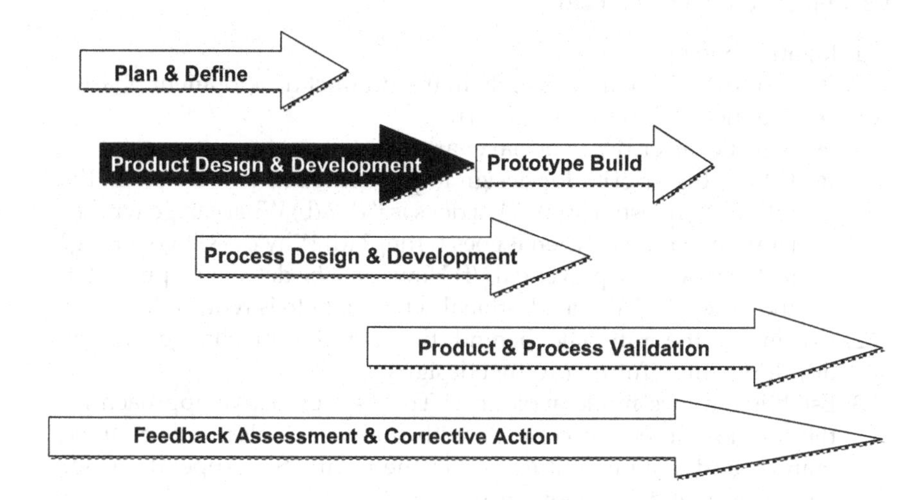

In this stage of the APQP we focus on four elements:

- Design FMEA
- DVP&R
- Subcontractor/Sub-supplier APQP Status
- Drawings and specifications

Directions

1. Form appropriate and applicable teams. The team must have ownership (authority and responsibility) to take "the" action required.
2. Each team will be assigned one of the four elements.
3. Each team will review the details of the assigned element and the appropriate checklist and present the results to the management team and/or the customer as required.

4. Be sure to address the following points in the presentation:
   a. Definition of the element
   b. Expectations of team members for the element
   c. When the element occurs
   d. Highlights of the element checklist

DFMEA

A group of activities intended to:

1. Identify potential:
   - Product failure modes early in the product development phase
   - Product design safety concerns
   - Significant/critical/special characteristics
   - Establish a priority for design improvement actions based on the following questions: (a) What does is it do? (b) What can go wrong? (c) What happens when it goes wrong? (d) Why does it go wrong? (e) How can it be prevented? (f) How can it be detected if present in the design? (g) What additional action or help is required?
2. Document the rationale behind product design changes to the development of future product designs.
3. Establish failure mode avoidance: This is a proactive approach for the engineer or team to come up with analytical tools in a disciplined manner and document it for use in the future. See Appendix D for more information on avoidance.

DFMEA expectations are as follows:

- Completing DFMEAs that focus on what has changed in the design and its effect on the entire system
- Creating a list of potential special characteristics (YC, YS) resulting from DFMEAs or equivalent analysis
- Addressing lessons learned and quality issues found in element CIR
- Ensuring that the DFMEA info is sent to PFMEA teams, including the potential YC and YS list

DVP&R

- Lists the engineering evaluations and tests required to establish that a design is fit for use in the intended environment,
- assists product development (PD) in verifying and documenting the functionality, reliability, and durability of the vehicle's design through planned tests and evaluations,
- documents the plans and reports for design verification of a given design, and

- identifies tests, acceptance criteria, sample sizes, and completion dates in a standard format

DVP&R expectations

1. Corrective action plans must be completed for any test result that does not meet design spec and reliability goals.
2. The DVP&R must:
   a. Align with the manufacturing and or customer agreed upon deliverables
   b. Provide inputs to complete reports and pro formas
   c. Identify any specific tests, methods, equipment, acceptance criteria, sample sizes, design level, and timing

Subcontractor APQP status

- The subcontractor APQP status element identifies and reports on the condition of external contractors' (i.e., suppliers) and subcontractors' APQP process.
- Internal and external suppliers identify APQP requirements by cascading program target APQP requirements to their subcontractors and by conducting APQP reviews, as appropriate.

Subcontractor APQP expectations
Subcontractor APQP status general expectations are that all internal and external suppliers must:

- Assess the risk and specify the level of their subcontractors' APQP participation
- Cascade APQP requirements to their subcontractors
- Hold regularly scheduled APQP review meetings with their subcontractors

Drawing and specifications

This element provides an evaluation of the development of targets and specifications to be input to the design process and drawings. These drawings include all engineering drawings, CAD data, material specifications, and engineering specifications.

Drawing and specifications expectations

- Program need dates (PND) are communicated and agreed to by affected areas.
- SCs and CCs are identified on the drawings or attached documents. (In some organization the identification may be

through other documentation; for example, with Ford Motor Co, you will find them in the Special Characteristic Communication Approval Form (SCCAF). Generally, they are test requirements that are important to customer satisfaction. As such, they are often product characteristic(s) for Ford components. All SCCAF items are required to be approved prior to PPAP. The signed copy of the SCCAF and the PPAP are required to be submitted to the product development and/or the supplier quality engineer. The SCCAF is always generated by the product development center of the customer and/or the supplier.
- Tolerances and part specs are compatible with accepted manufacturing standards for gauges and equipment. Gauges must be able to measure 10× the specification measurement of the product.

Drawings and specifications must include:

- Engineering specification tests.
- Product validation test requirements.
- Documentation to support prototype build and pre-launch control plan development.
- The drawing information and engineering specifications that will be used as a prerequisite to the process-based business collaboration platform (PBCP).
- The material used by the suppliers or the subcontractors must be approved and included in the customer's source list.
- PD and manufacturing engineering (ME) personnel will assess drawings and specifications to ensure they meet all of the program's quality requirements.

Prototype build

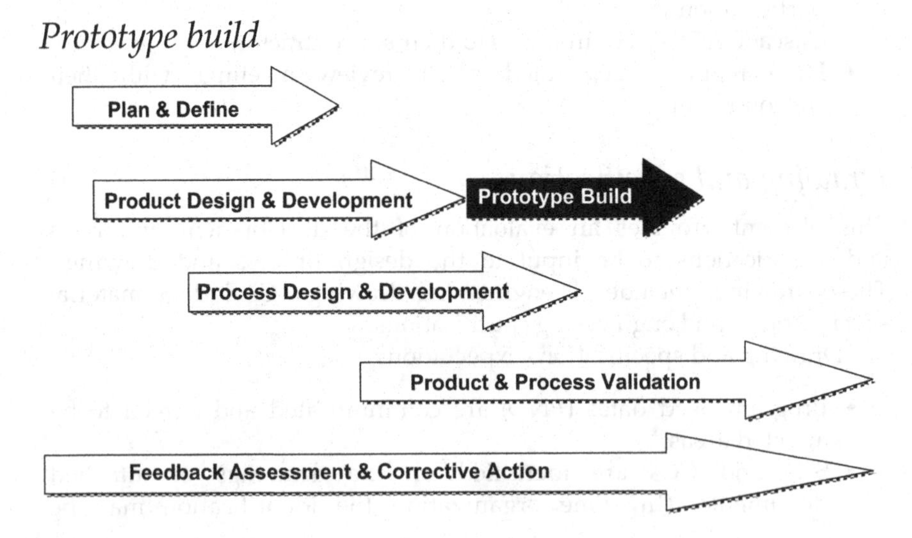

Prototype build overview

Prototype build defines the manufacture and assembly of components, subsystems, systems, and vehicles supplied to the customer for builds occurring prior to the 1PP trial run. It also includes not only the actual build process, but also the preparation for the build so that CPs are supplied to the customer prior to 1PP.

APQP prototype build elements support the successful planning and construction of prototype build units, which provide an understanding of the vehicle's performance:

- Design
- Functionality
- Manufacturability

Prototype build elements

The prototype build control plan (PBCP) supports the planning and construction of prototypes by documenting the process steps of the Prototype Build phase. The Prototype Build ensures that the customer receives control plans (CPs) manufactured and assembled prior to 1PP units (at assembly plant).

PBCP definition

1. The PBCP description of the dimensional measurements and material/functional tests that occur during the CP build.
2. It is the first summary document that indicates:
    - The steps between prototype build and pre-launch leading to the production control plan.
    - The process steps to identify and align both the product's characteristics and its targets.

PBCP general expectations

1. Product characteristics are reviewed.
    a. Characteristics required for PBCPs are identified.
    b. All potential significant (YS) and critical (YC) characteristics from the DFMEA's must be included.
2. The process parameter conditions must be documented. This includes all special tooling, manufacturing, and assembly conditions.

Prototype build expectations

- PBCPs are followed.
- Suppliers receive all necessary information and participate according to the program's schedule and timing.
- The timing for changes will be in line with the next build phase.

- All prototype materials must meet 100 percent of the dimensional specifications and functional requirements at delivery to meet the CP in-plant date.
- Quality levels are checked against the PBCP.
- Dimensional and functional data are available for review by the customer on each part.
- Suppliers provide customers with any testing and checking requirements for prove-out of their part design.
- An "action matrix" is developed to capture and resolve all concerns found during build and any subsequent lab, plant, and/or road testing, and all concerns are reviewed with the program team and appropriate management.

chapter three

Process design and development

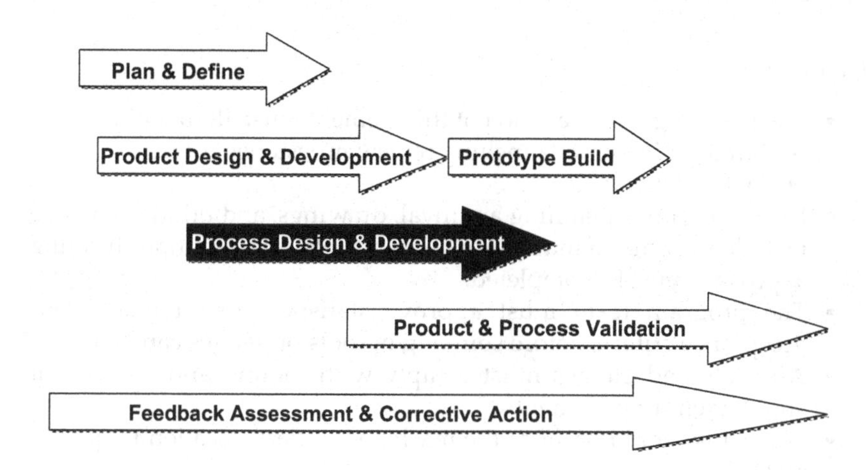

Objectives

1. List the seven process design and development elements.
2. Describe where each process design and development element fits in the APQP process.
3. Describe the following items for each process design and development element:
   a. Definition
   b. Expectations
   c. Questions on checklists

Elements

1. Facilities, tools, and gauges
2. Manufacturing process flow
3. Process FMEA
4. Measurement system evaluation
5. Pre-launch control plan
6. Operator process instructions
7. Packaging specifications

Facilities, tools, and gauges

This element identifies the new, additional, refurbished, and relocated facilities and resources necessary to manufacture the customer-specified product at the designated quantity and quality levels. It also identifies the tools and gauges used in process machinery to transform raw material into finished parts or assemblies.

Expectations

- The planning and execution of this element must align with:
    - Strategic intent to launching review milestones
    - PQOS deliverables
- Facility permits, planning approval, drawings, and utilities must be included on the manufacturing timing plan. In addition, funding approvals must be completed.
- The program team must approve statistical requirements and acceptance criteria before sourcing of tools or gauges can be made.
- All tools and gauges must comply with tooling and equipment supplement requirements.
- Trial runs should occur at the machine builder's location to qualify all tools.
- All corrective actions for facilities, tools, and gauges not meeting customer requirements must be completed before the production trial run.
- Tools and gauges must be delivered, set up, and approved in their specified facilities before the production trial run.

Manufacturing process flow

This element is a graphic representation of the current or proposed sequence of the manufacturing process to ensure that the process definition, PFMEA, and CPs can be created and analyzed in the appropriate sequence.

Manufacturing process flow expectations

Manufacturing process flow general expectations are that:

- A cross-functional team led by Vehicle Operations (VO) Engineering develops a manufacturing process flowchart to serve as an input to PFMEAs.
- The process flowcharts are developed using PFMEAs and surrogate processes to serve as input for new processes and technologies.
- The current production flowcharts may be modified to show revisions to the process flow.

PFMEA

The PFMEA is a systematic approach used by a manufacturing-responsible team to ensure that potential process-related failure modes and their associated causes have been considered and resolved. It is always linked to the DFMEA, process flow diagram and CP.

PFMEA expectations

1. All PFMEAs are prepared by a cross-functional team led by ME following the steps defined in the FMEA guideline.
2. Process FMEAs must be established any time that:
   a. New processes, technologies, products, or product features are introduced.
   b. Quality risks are identified that cannot be resolved through design changes.
   c. Major quality concerns have not been resolved during the current production model year.
   d. List of "recommended actions" and lessons learned is incorporated in the process (continuous improvement).
   e. Mistake proofing has been used in addressing corrective actions—especially in inspection-dependent processes.

Measuring system evaluation

This element assesses the variation of the measurement system and determines whether it is acceptable for monitoring the production process.

Expectations

- Correlation standards are reached and agreed to with the customer for duplicate gauges or test equipment.
- Gauge instructions and visual aids to ensure appropriate use of gauge and test equipment in production.
- All measurement systems (gauges and test equipment) must be modified to reflect the latest engineering part level before the production trial run.
- The user/customer must be given the opportunity to review and concur with the gauges and test equipment study results before the production trial run.
- The measurement system evaluation must be repeated and approved following all gauge and test equipment modifications.

- The production control plan must align with the measurement systems development plan.
- Gauge repeatability and reproducibility studies must be conducted on all measurement systems.

Pre-launch control plan

This is a description of the dimensional measurements and material/functional in-process tests or checks that occur after prototype builds, but before full production.

Expectation

- Legitimate/appropriate inspection sample sizes, frequencies, control method, etc. inclusive to pre-launch phase of production are included.
- All product and process characteristics for pre-launch control plans are identified and reviewed.
- Results from the PBCP and DFMEA provide the outline for the pre-launch control plan.
- The production control plan for the current production model is used to assist in the development of the pre-launch control plan if APQP Elements 8 and 13 have not been initiated.

Operator process instructions

1. This element describes the details of the controls and actions that operating personnel must perform *currently* to produce quality products.
2. These instructions are divided into two major components:
    - Description of the process
    - Operator instructions

Expectations

- A completed operator process instructions package.
- Instructions developed to be used during a production trial run to ensure the process performs as intended.
- Operator process instructions describe all process steps necessary to produce a quality product. The training instructions must include:
    - Detailed instructions
    - Visual aids
    - Any other quality materials necessary to support the production operator

Packaging specifications

This element describes the design and development of the end-item shipment packaging, including interior partitions.

Expectations

- Past issues on similar or surrogate packaging are discussed and resolved by the team.
- Corrective actions exist for any packaging issues identified that affect product performance and characteristics.
- Supplier and the receiving plant agree upon packaging requirements.
- Packaging evaluation must test the packaging under the expected conditions of transport and material handling.
- Packaging design must ensure that the product performance and characteristics will remain unchanged during packing, shipping, and unpacking.

chapter four

Product and process validation

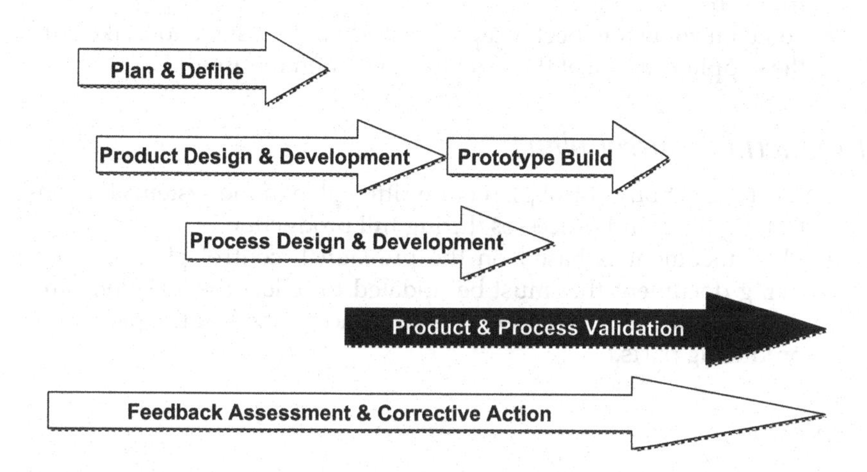

Elements

1. Production trial run
2. Production control plan
3. Preliminary process capability
4. Production validation testing
5. Part submission warrant (PSW)

Production trial run

The production trial run is a validation of the effectiveness of the manufacturing and assembly process, using production tooling, equipment, environment (including production operators), facilities, and cycle times.

Expectations

- Production facility, tooling, equipment, personnel, packaging, operators, PSW-1 parts, and pre-launch control plans are used on the trial run.
- Concerns from the trial run that could prevent run at rate and capacity by date required are listed.

- Corrective design and process actions are established.
- The pre-launch control plan is followed during the 1PP production trial run.
- The production trial run is used to confirm or add alignment between product and process characteristics.
- Operator process instructions are followed during the production trial run.
- Some customers expect a capacity study at this stage to make sure the supplier can meet the weekly production required.

Production control plan

- The production control plan is a written plan of the systems for controlling parts and processes during full production.
- This document is based on the pre-launch control plan and is a living document that must be updated to reflect the additions and deletions of controls, based on experience gained in the process of producing parts.

Expectations

- Reaction plans specify corrective and containment actions necessary to avoid operating out-of-control or producing nonconforming products.
- The production control plan is developed by a cross-functional team led by ME.
- The production control plan follows the methodology documented in the AIAG APQP and control plan reference manual.
- The outcomes of the final PBCP provide the basis for the pre-launch control plan that in turn becomes the source for the production control plan.

Preliminary process capability

The preliminary process capability is a statistical study and assessment of the ability to produce product within specifications.

Expectations

- Results are analyzed to verify statistical stability as well as capability. Normality must be established at this stage, otherwise the capability indices (C_{pk}, P_{pk}, etc.) will not make sense.
- The statistical and analytical techniques used to determine capability must be acceptable to the customer.

- Preliminary process capability studies must be performed in a way that aligns with the development of the pre-launch control plan.
- Preliminary process capability studies must be completed, which means that the customer must have the opportunity to review the studies before production part approval.

Production validation testing

- Engineering tests validate that products made from the production tools and processes meet engineering specifications.
- Internal and external suppliers to the customer must complete production validation (PV) testing as a requirement of a PSW for 1PP.

Expectations

- A PV test plan, including progress achieved, is established.
- Concerns identified during PV testing are assigned, resolutions are implemented, and solutions are tracked through closure of the concern.
- All parts, end-items or components, subsystems, and systems for PV testing must be completed before the FEU build according to the engineering specifications in the pre-launch control plan.
- All customer-specified dimensional, material, functional, and reliability tests must be completed before production part approval.

Part submission warrant (PSW)

- The PSW ensures that engineering design requirements are met by the supplier and that the process used is capable of producing parts that meet these requirements.
- The PSW is the final sign-off verifying that the PPAP has been followed.

Expectations

- Concerns identified during PPAP and PSW have been resolved.
- The customer's material in plant date (IPD) must be included in the supplier's timing plan.
- Production part approval requirements must be completed prior to the IPD of the user plant.
- All items of the AIAG PPAP manual must be completed and the required documentation provided to the customer with the PSW.

Status reporting

- Design and manufacturing reviews
- This element describes the activities that are necessary to make the review meetings effective and:
  - Prevent problems and misunderstandings
  - Provide a mechanism to monitor the program's progress and resolutions
  - Review the product and manufacturing process detail (e.g., part drawings)
  - Provide data-driven verification of designs and processes

Expectations

- Review APQP status reports during the design reviews and discuss relevant issues.
- Resolve design feasibility concerns in time to support each build's IPD/material review date (MRD).
- Appraise the progress of the design verification plan (unanticipated failure modes encountered during design verification testing must be addressed in the DFMEA).
- Assess the progress toward achieving reliability, quality, cost, and timing targets.

chapter five

Corrective and preventive action feedback

Problems occur in everything we do. The way we handle them speak volumes about our attitude in quality matters. Fundamentally, there are two ways to deal with or handle problems: the first one is to have a corrective approach, and the second is to have a prevention strategy.

The corrective approach has many methodologies and tools that may be used. Some of them are indicative of the core business strategy of an organization—see Figure 5.1. However, the corrective approach is limited because it only focuses on the "fix" for a given problem. That is where prevention becomes imperative and necessary. We must plan so that the problem we just fixed will not appear again.

Perhaps one of the most common methodologies to solve a given problem (corrective action) is the embellished scientific method, known as the 8D method. The 8D approach is the most beneficial under the following six criteria:

1. The symptom has been defined and quantified.
2. The G8D customer(s) who experienced the symptom(s)—and when appropriate, the affected parties—have been identified.
3. Measurements taken to quantify the symptom(s) demonstrate that a performance gap exists, **AND/OR** the priority (severity, urgency, growth) of the symptom(s) warrants initiation of the process. Note that this priority is not the same as the one used for FMEA (severity, severity × occurrence, × detection—[RPN]).
4. The cause is unknown.
5. Management is committed to dedicating the necessary resources to fix the problem at the root level cause and to prevent recurrence.
6. The complexity of the symptom exceeds the ability of one person to resolve the problem.

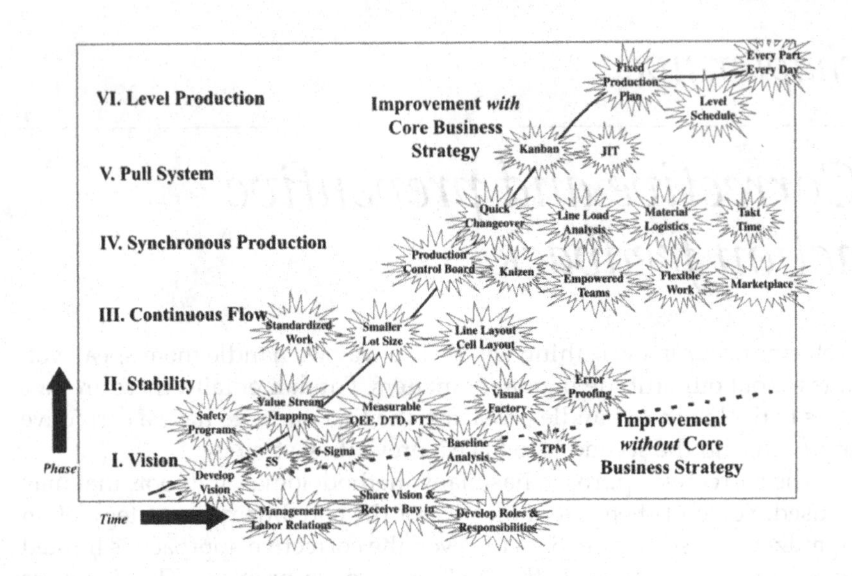

Figure 5.1 Core business strategy.

Summary of the 8D method

D0—Prepare for the G8D process

1. Evaluate the need for G8D process.
2. Provide emergency response action, if needed.
3. Use the G8D application criteria to determine whether to use the G8D process.
4. Use the assessing questions at each step in the G8D process as a formative process and check a confirmation of readiness.

Typical D0 tools

- Trend chart: displays performance over time
- Pareto chart: shows ranking based on frequency
- Paynter chart: shows effectiveness (is it working?) of corrective actions

D1 summary—establish the team

1. Team memberships
2. Team roles
3. Operating procedures
4. Team system model
5. Team synergy

Typical considerations for D1

- Establish a small group of people with the procedures and/or product knowledge. (They must have ownership of the process and or problem under consideration.)
- Establish the allocated time, authority, and skills required in technical disciplines to solve the problem and implement corrective actions.
- Define success for the team.
- Be clear on the commitment of both management and the team.
- Establish the composition of the team as a cross-functional and multidiscipline.
- Make sure there is a designated champion and team leader.
- Allow for the development of the team-building process.

D2 summary—describe the problem

1. Problem statement: objects and defect
   a. Problem must be stated in a simple, concise way that identifies the object and defect of a problem for which the cause is unknown.
   b. Statement is derived by using the process question "What's wrong with that?" and testing it with repeated whys.
2. Problem description:
   a. The output of a process used to amplify the problem statement.
   b. Defines the boundaries of a problem.
   c. Narrows the search. Describes the events in terms of the is/is not analysis: what, where, when, and how big.

Common tools used in D2

- Process flow diagram: displays where in the process the problem is occurring.
- Cause and effect diagram: identifies the universe of possible causes.
- Subdivide the problem: deal with one problem statement at a time.
- Observations versus conclusions: be aware of what side you are working from.

D3 summary—Develop interim containment action (ICA)

1. Verification is proof *before* implementation.
2. Validation is proof after implementation.
3. ICA: *it is a temporary fix* and is kept in place until verified permanent corrective action (PCA) can be implemented.

D4 summary—define and verify root cause and escape point

1. Root cause: a verified cause that accounts for the problem; verified passively and/or actively, by making the problem come and go on demand.
2. Escape point: the earliest location in the process, closest to the root cause, where the problem could have been detected, but was not.
3. Verification of the root cause and escape point.
4. Never-been-there: a new, higher level of performance desired.
5. Change-induced situation: degradation, abrupt change and never been there. Things were once acceptable but now they are not.

D5—Choose and verify PCAs

Purpose

1. To select the best (it implies that there are others) PCA to remove the root cause.
2. To select the best PCA to address the escape point.
3. To verify that both decisions will be successful when implemented.

Rationale

1. Making the best decision
2. Evaluating benefits and risks
3. Not rushing to implementation (make sure you're not implementing ICA for a quick fix)
4. Verify the choice (the fix) will work
5. Do not cause undesirable effects or other problems

Summary

1. It is critical to take the time to make the best decision regarding the PCA at D5.
2. Use the team decision-making process best suited to the specific needs of the team.

Decision-making process

There are several ways that we may arrive at a decision. The most common types are:

- Unilateral: this is the easiest to arrive at but the weakest of them all.
- Polling (voting).
- Prioritization (rational).
- Compromise (50 percent each; may not get 100 percent of each).
- Consensus (it is the only way to win in any resolution). This is the preferred method, but the most time consuming.

D6—Implement and validate PCAs

Purpose

1. To plan and implement selected PCA(s)
2. To remove the ICA
3. To validate the PCA (show me)
4. To monitor the long-term results

Planning and problem prevention

1. What could make the step/phase go wrong?
2. What can be done to prevent that from creating the problem?
3. What needs to be done if it happens anyway?
4. Who needs to initiate the contingent action, and what information should tell them to proceed?

D7—Implement recurrence prevention

Purpose

1. Modify the necessary systems including policies, practices, and procedures to prevent recurrence of this problem and similar ones.
2. Make recommendations for systemic improvements, as necessary.

Rationale

1. Fixes the root cause of the root cause of the problem.
2. Addresses the systems, practices (what fixed the issue), policies (for documentation purposes), and procedures that allowed the problem to occur and escape.

Recurrence prevention is

Action taken to prevent the recurrence of the present problem, similar problems, and systemic problems. Corrective action is incomplete without preventive actions!

Summary

1. Behind all root causes are systematic issues that need to be fixed.
2. The goal is to change the system that allowed the problem to occur in the first place.
3. D7 provides the opportunity to prevent similar problems from happening again.
4. The team also has the opportunity to offer recommendations for systematic improvements.

D8 summary—Recognize team and individual contributions

1. Recognize individual and team contributions.
2. Complete any unfinished team business before disbanding team. Remember to *always* congratulate in public but criticize in private!
3. Celebrate the team's experience.

5D

The 5D approach to problem solving is a shorter version of the 8D and is used generally as a summary for the management team during product launches. Generally, the methodology is used as a "possible" problem solving methodology, since during launch one may not have all pertinent information. The five steps are:

1. Issue description (Verb-noun)
2. Containment (Make sure the customer does not receive a non-conforming part)
3. Root cause (What is the potential root cause? Develop a work plan—strategy, tactics and tools. Try to isolate the escape point).
4. Corrective action (What is the potential corrective action?)
5. Preventive action (What are the plans to prevent the non-conformance from happening again or spreading in similar situations?)

5 Whys

5 Whys is an iterative interrogative technique used to explore the cause-and-effect relationships underlying a particular problem. The primary goal of the technique is to determine the root cause of a defect or problem by repeating the question "Why?" Each answer forms the basis of the next question. The "5" in the name derives from an anecdotal observation on the number of iterations needed to resolve the problem. In reality, there may be a need for less or more than five "whys."

Not all problems have a single root cause. If one wishes to uncover multiple root causes, the method must be repeated asking a different sequence of questions each time. The method provides no hard and fast rules about what line of questioning to explore, or how long to continue the search for additional root causes. Thus, even when the method is closely followed, the outcome still depends upon the knowledge and persistence of the people involved. In order to carry out the *5-Why* analysis properly, it is important to follow these steps:

1. It is necessary to engage the management in the company in the 5-Why process. For the analysis itself, consider what the right working group is, and also consider bringing in a facilitator for more difficult topics.

2. Use paper or a whiteboard instead of computers.
3. Write down the problem and make sure that all people understand it.
4. Distinguish causes from symptoms.
5. Pay attention to the logic of the cause-and-effect relationship.
6. Make sure that root causes certainly lead to the mistake by reversing the sentences created as a result of the analysis with the use of the expression "and therefore."
7. Try to make the answers more precise.
8. Look for the cause step by step: don't jump to conclusions.
9. Base the statements on facts and knowledge.
10. Assess the process, not people.
11. Never leave "human error," "worker's inattention," "blame John," etc., as the root cause.
12. Foster an atmosphere of trust and sincerity.
13. Ask the question "Why" until the root cause is determined, i.e., the cause the elimination of which would prevent the error from occurring again.
14. When you form the answer for the question "Why," it should happen from the customer's point of view.

Benefits of the 5 Whys

- Help identify the root cause of a problem.
- Determine the relationship between different root causes of a problem.
- One of the simplest tools; easy to complete without statistical analysis.

When is 5 Whys most useful?

- When problems involve human factors or interactions.
- In day-to-day business life; can be used within or without a Six Sigma project.

How to complete the 5 Whys

1. Write down the specific problem. Writing the issue helps you formalize the problem and describe it completely. It also helps a team focus on the same problem.
2. Ask Why the problem happens and write the answer down below the problem.
3. If the answer you just provided doesn't identify the root cause of the problem that you wrote down in Step 1, ask Why again and write that answer down.
4. Loop back to step 3 until the team is in agreement that the problem's root cause is identified. Again, this may take fewer or more times than five Whys.

Special note: While the 5 Whys is a powerful tool for highly technical individuals and complex problems, there is a built-in flow in its simplicity for solving serious problems and arriving at the root cause. That is, the method is too basic for analyzing a root cause. Some of the problems that the investigator(s) or analyst(s) must be aware of and consider are:

- Tendency for investigators to stop at symptoms rather than going on to lower-level root causes.
- Inability to go beyond the investigator's current knowledge—cannot find causes that they do not already know.
- Lack of support to help the investigator ask the right "why" questions.
- Results are not repeatable—different people using 5 Whys come up with different causes for the same problem.
- Tendency to isolate a single root cause, whereas each question could elicit many different root causes.

These can be significant problems when the method is applied through deduction only. On-the-spot verification of the answer to the current "why" question before proceeding to the next is recommended to avoid these issues. In addition, performing logical tests for necessity and sufficiency at each level can help avoid the selection of spurious causes and promote the consideration of multiple root causes.

3 × 5 Why

To avoid some of the issues that the 5 Why methodology presents, the 3 × 5 Why may be used—see the form in Figure 5.2. Fundamentally this methodology presents three separate steps of 5 Whys, each representing a unique opportunity to solve a particular problem. They are:

1. *Specific step or leg* focusing on the specific "problem" by asking as specifically as possible: "Why did this specific situation happen?"
2. *Detection step or leg* focusing on the escape point of the problem by asking as specifically as possible: "Why was this situation overlooked?" This is a profound question because it forces the analyst to think about what could have been in place to prevent the nonconformance.
3. *Systemic step or leg* focusing on why the possibility existed for the nonconformance to exist. This step is interested in identifying the weak spot of the system. Generally, this is a management issue, and it is this reason that most problems are not fixed. Analysts are intimidated by management therefore they do not reveal to management the real problem(s) which they are causing. Generally, these problems are systemic and are controlled by management.

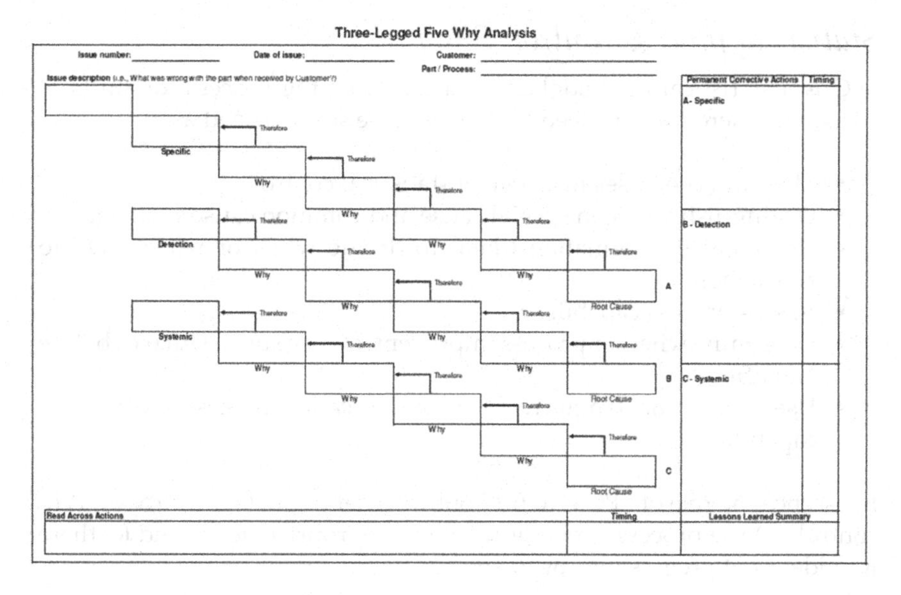

***Figure* 5.2** 3 × 5 Why form.

The 5 Whys is an iterative interrogative technique used to explore the cause-and-effect relationships underlying a particular problem. The primary goal of the technique is to determine the root cause of a defect or problem by repeating the question "Why?" Each answer forms the basis of the next question. The new procedure, 3 × 5 Why, creates three lines of inquiry. The "Why, Why, Why …" process is applied independently to each of the following legs:

1. Specific Leg: Why did this specific situation happen? This is the normal non-conformance being analyzed using the 5-Why Procedure.
2. Detection Leg: Why was this situation overlooked? This asks why our in-place detection procedures didn't catch this.
3. Systemic Leg: Why did the possibility exist for this situation to occur? This asks you to look at what about the larger organization, systems, or procedures creates an environment in which this nonconformance occurred.

Remember your customer wants to be protected from your nonconformance and has an interest in you evolving your systems: that is the purpose of the 3 × 5 W requirement. Meet this purpose and everyone should be happy regardless if you use detect and sort or the more powerful elimination of variance from within your manufacturing process.

Statistical process control (SPC)

SPC is a statistical methodology that monitors the process for shifts of variation. Therefore, the need for SPC may be summarized as:

- Maintain processes in a state of statistical control
- Distinguish between special cause and common cause variation
- Use reliable information (data) to decide when or when NOT to take action
- Assess process capability
- Determine whether process improvements are successful (i.e., before and after)
- Use a common language for communicating process control and capability

In essence the role of SPC is to find out whether or not (a) the process is in control, (b) the process is operating at the appropriate level, and (c) there is evidence of process improvement.

Operational requirements

- Comprehension, awareness and commitment at the highest level
- Use as a common language
- Creation of an environment free of fear
- Recognition that total involvement is an essential ingredient of success
- Appreciation of need for "constancy of purpose"
- Process ownership is established and appropriate process parameters/ product characteristics are identified
- Creation of a system whereby operators (in particular) are fully involved in any problem-solving effort

Implementation guidelines

- Management commitment
- Steering group
- SPC Co-ordinator/Facilitator
- Training
- Pilot scheme
- Benchmark
- Installation
- Review and expansion

Process capability requirements

1. Process must be in control
2. Need to know:
   a. Specification limits
   b. Mean of process output population, μ_x
   c. Standard deviation of process output population, σ_x
   d. Shape of process output distribution—we expect a normal distribution. With a normal process we can be sure of stability, consistence and predictability

Basic steps

To conduct an effective SPC there are seven steps essential for success. They are shown in Figure 5.3.

Basic process

The basic process is shown in Figure 5.4

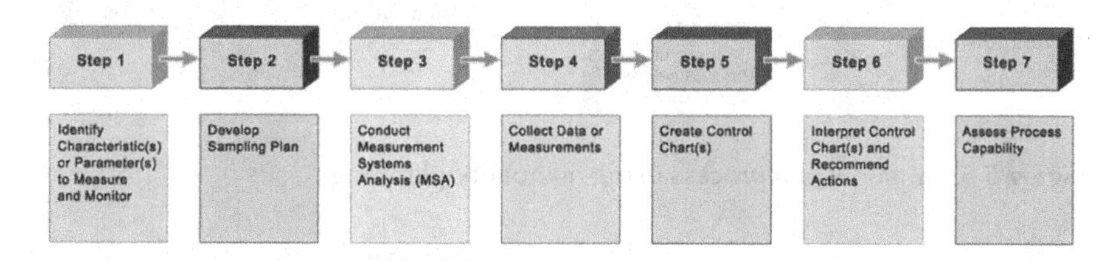

Figure 5.3 The seven essential steps to SPC.

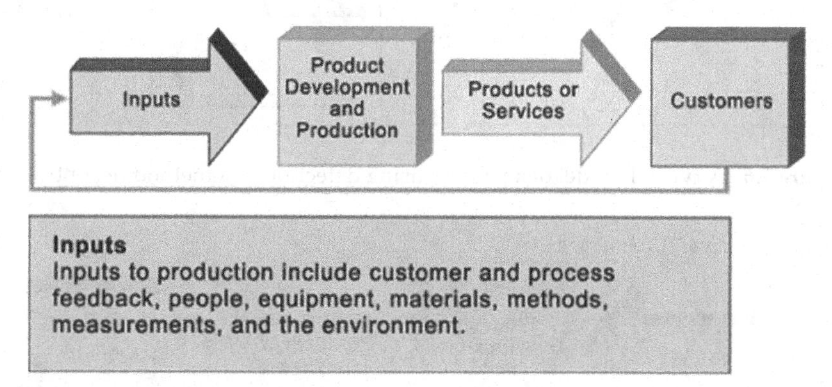

Figure 5.4 A typical basic process.

Consequences of no quality

A typical model of a process not using quality planning, i.e., SPC Figure 5.5.

Detection strategy

A typical model of a process using detection as a method of control is shown in Figure 5.6.

Prevention strategy

A typical model of a process using prevention as a method of control is shown in Figure 5.7. For a detailed discussion on SPC see Stamatis (2002); AIAG:SPC (2005); Wheeler (2010).

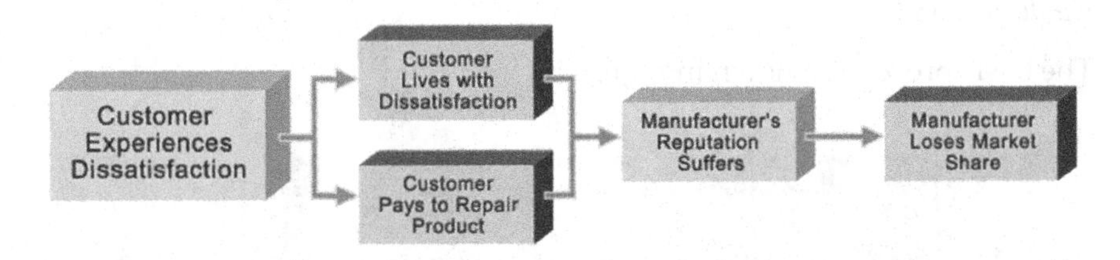

Figure 5.5 A model of a process using no quality planning.

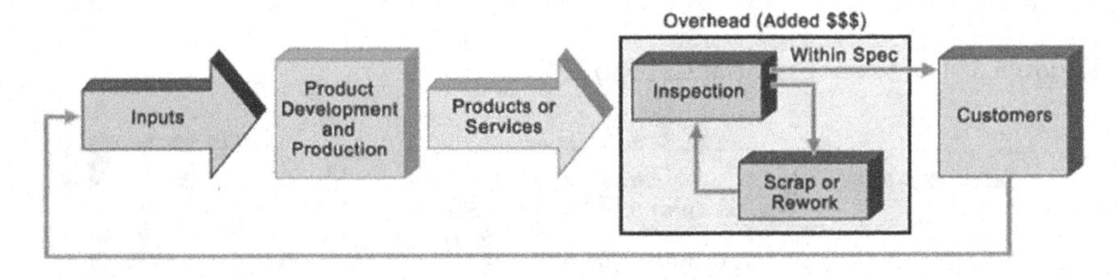

Figure 5.6 A typical model of a process using detection as a method of control.

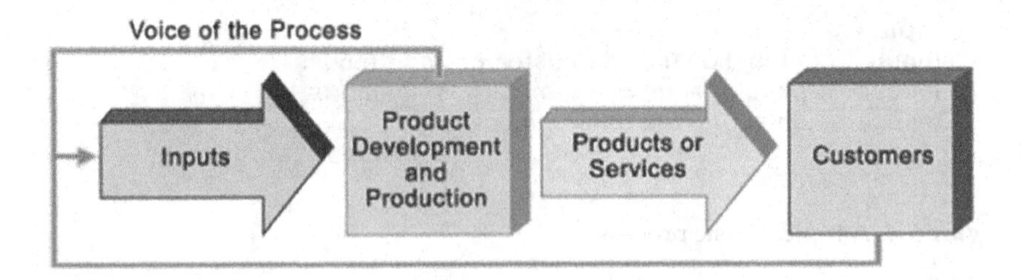

Figure 5.7 A typical model of a process using prevention as a method of control.

The Six Sigma methodology

Six Sigma is a methodology for process improvement. It can be applied to manufacturing (technical) and business (transactional) processes to improve the design and/or manufacturing of products, services, and manufacturing processes. Six Sigma is not a set of "all new" tools and processes; rather, Six Sigma provides an overall approach for using existing tools to reduce defects and increase customer satisfaction. There are two approaches: (a) the DMAIC model and (b) the DCOV model (DFSS). For a very detailed discussion of the Six Sigma methodology, see Stamatis (2001–2002); and Breyfogle (2003).

What makes Six Sigma special?

- Leadership commitment, competence & involvement
- Methodology & tools
- Data driven
- Statistically validated
- Best people 100 percent dedicated to defect reduction
- Project focused

Key points of the methodology

- Assign best people—100 percent dedication—to solve biggest customer satisfaction problems
- Link customer satisfaction with financial results: an integrated business strategy
- Integrate existing tools into a coordinated deployment system
- Data driven, results driven
- Can produce quick short termed results that build momentum for sustained gain
- Provides leadership training for new leaders
- Management commitment

DMAIC (Define, Measure, Analyze, Improve, Control)

DMAIC is a systematic methodology for reducing defects in any manufacturing or service process. Specifically, the DMAIC model focuses on correcting existing problems and preventing the recurrence of such problems. Typically, these problems are associated with the manufacturing process, and this model is best suited for assignable cause problems. Each letter in the DMAIC acronym represents a process phase:

- Define: A problem is defined and a potential Six Sigma DMAIC project is scoped out.
- Measure: Data on current process performance is measured.

Table 5.1 The relationship of SPC and the DMAIC model

Define	SPC is used to determine the magnitude of the problem and the baseline capability
Measure	SPC is used to determine the magnitude of the problem and the baseline capability.
Analyze	
Improve	SPC is used to confirm the success of the improvement.
Control	SPC is used to monitor the process improvement and stability and to confirm process capability.
Replicate	[Usually replication is added to make sure that the actions taken can be replicated. It is not a formal step of the DMAIC, but it is a good step to follow.]

- Analyze: Data is analyzed to determine the source of the problem and how the process might be improved to eliminate the problem.
- Improve: An improvement option is selected and implemented.
- Control: Ongoing performance is monitored and controlled.

The DMAIC model may be used with SPC and that relationship is shown in Table 5.1. Table 5.2 displays some common tools used in the DMAIC model.

DCOV (Define, Characterize, Optimize, Verify)

DCOV is a disciplined methodology for designing products and processes that perform at or above customer expectations. DCOV is the basic and common model for the broader concept of Design for Six Sigma (DFSS). Specifically, the DCOV model focuses on correcting anticipated problems and preventing the recurrence of such problems. Typically, these problems are associated with the design and or manufacturing process, and this model is best suited for common cause problems. Each letter in the DCOV acronym represents a design phase:

- Define: Customer requirements and wants are synthesized into product and/or process characteristics.
- Characterize: Product and manufacturing process design concepts are generated and selected.
- Optimize: The product and its manufacturing process are designed to be robust and reliable.
- Verify: The expected performance level of the new process or product is confirmed.

Table 5.2 Typical tools used in the DMAIC model

Common tools used in the DMAIC model				
Define	Measure	Analyze	Improve	Control
• QFD surveys • GQRS • Warranty • In-plant data • Pareto	• Blueprints • Process • Sheets • Metrics • Flow charts • Linkage matrices • FMEAs • Measurement system analysis • Capability analysis	• Multivariate analysis • SPC & charting basic stats • Hypothesis • Basic DOE • Check sheets • Histograms • Scatter diagrams • Regression	• Designed experiments • Full factorial and fractional comparison • Measurement system analysis • Capability statistical tolerancing of process	• Control plan • Operator illustrations • Operator instructions • Error Proofing • Automated control • QPS • Variation monitoring

When to use DFSS

- *New* product or service introduction
- Process *broken* or *does not exist*
- Process has reached *entitlement,* which is when the process has only random variation and improvement must be undertaken

Right questions for management

1. What processes (activities) are you responsible for?
2. What processes have the highest priority for improvement? How did you come to this conclusion? Where is the data that led to this conclusion? *COPQ, rolled throughput yield.*
3. How is the process performed? *Process map.*
4. What are your process performance measures? Why? How accurate and precise is your measurement system? *GR&R, ANOVA.*
5. What are the customer-driven specifications for all of your performance measures? How good or bad is the current performance? Show me the data. What are the improvement goals for your process? *QFD, process capability.*
6. What are all the sources of variability in the process? *C&E, FMEA.*
7. Which sources of variability do you control? How do you control them and what is your method of documentation? *C/Ns, SOPs, control plans, SPC.*
8. Are any of the sources of variability supplier dependent? If so, what are they, who is the supplier, and what are we doing about it?
9. What are the key variables that affect the average and variation of the measures of performance. How do you know this? Show me the data. *ANOVA, ANOM, F test, T test, Chi square, DoE.*
10. What are the relationships between the measures of performance and the key variables? Do any key variables interact? How do you know for sure? Show me the data. *DOE, regression.*
11. What setting for the key variables will optimize the measures of performance? How do you know this? Show me the data. *Optimization routines.*

SPC used in DCOV

There is a tremendous misconception in the industrial world that corrective action is sufficient to improve quality. They forget that fixing problems always occurs after something has gone wrong. Obviously, one must fix the problem, but unless it is avoided or prevented from happening again, many people will be busy but not improving.

The focus of quality should be on prevention (before things happen) rather than detection (after they happen). The reason for this is that it is more

Table 5.3 The relationship of SPC and the DCOV model

Define	SPC may be used to determine the problem's magnitude in an existing process and the baseline capability for an external or internal product benchmark.
Characterize	SPC may be used to assess capability of an existing process. This process may be a surrogate for the future process
Optimize	SPC may be used to confirm the success of the improvement
Verify	SPC may be used to verify long-term process capability and to continuously look for changes in the process.

economical and efficient to do things "right" the first time rather than do them all over again. What happens, however, most organizations act on the convenience and expediency of short-term earnings, or to paraphrase Deming, as long as we focus on the "quarterly earnings, true quality will suffer."

There are many ways an organization can emphasize prevention. However, in this chapter we summarize the DCOV Six Sigma model. The relationship of SPC and the DCOV model is shown in Table 5.3.

Corrective action preventive action (CAPA)

In the health and pharmaceutical industries, a specific corrective and prevention methodology is used to meet the government requirements of FDA 21 CFR 820.100, and this methodology is known as corrective action and preventive action (CAPA). (Of note is that it is mentioned in the Compliance of Good Manufacturing Practices [CGMPs] a total of 15 times and it is inferred a host of others). When failure does occur, we demand a rigorous process of investigation be initiated to identify why it occurred. CAPA is a process that investigates and solves problems, identifies causes, takes corrective action, and prevents recurrence of the root causes. The ultimate purpose of CAPA is to assure the problem can never be experienced again. CAPA can be applied in many disciplines. A few of these disciplines are:

- manufacturing;
- product design;
- testing verification and validation;
- distribution, shipping, transport and packaging; and
- use-applications.

CAPA is the result of a US FDA requirement, FDA 21 CFR 820.100. The CAPA requirement applies to manufacturers of medical devices and compels them to include CAPA in their quality management system (QMS). CAPA is split between two distinct but related functions. For a good application of the CAPA approach see Durivage (2017).

1. Corrective action (CA) is an extension of root cause analysis (RCA). The first goal of CA is to find the root cause, base event, or error that preceded the problem. The second goal is to take action directed at the root cause or error.
2. Preventive action (PA) is similar to lessons learned/read across. PA resembles the replication activity of DFSS. Another example of PA in industry is *Yokaten,* a Japanese term used by Toyota, describing a sharing across the organization. The primary goal of PA is to inform an organization and prevent the problem from returning in other facilities, lines, or products.

Why implement CAPA?

Identifying the root cause of failure is a key tenet of any effective QMS. When a problem occurs, it is often just a symptom of the real issue. Symptoms can be treated, but finding out why the symptom is experienced is the true purpose for implementing CAPA. Failure to implement an effective CAPA process is a violation of FDA regulations defining good manufacturing practice (GMP). Once implemented, the CAPA system must exhibit ten objectives to meet the intent of the FDA 21 CFR 820.100 requirement (for more information see (CAPA 2018); The Lean Enterprise, http://www.freeleansite.com/). The ten objectives of CAPA implementation are:

1. Verification of CAPA system procedure(s) that addresses the requirements of the quality system regulation. It must be defined and documented.
2. Evidence that appropriate sources of product and quality problems have been identified.
3. Tracking of trends (which are unfavorable) are identified.
4. Data sources for CAPA are of appropriate quality and content.
5. Verify that appropriate statistical process control (SPC) methods are used to detect recurring quality problems.
6. Verify the RCA work performed is aligned to the level of risk the problem has been identified with.
7. Actions address the root cause and preventive opportunities.
8. CAPA process actions are effective and verified or validated prior to implementation.

9. CAPA for product and quality problems are implemented and documented.
10. Nonconforming product, quality problems and corrective/preventive actions have been properly shared and included in management review.

How to implement CAPA

There are many ways to apply the two functions of CAPA. The two basic ones are: (a) CA and (b) PA approach:

Corrective action

When a symptom is observed or communicated, a systematic set of activities is initiated. The activities are intended to describe the problem in sufficient detail so that the team can identify a root cause path. Once a root cause path is selected, a permanent CA is identified, verified, implemented, and validated. The Quality-One nine steps for CA are detailed below:

1. Symptom is observed or communicated. The symptom must be quantified through the application of five questions, or 5Q (5 Why), and confirmed as a true symptom, worthy of defining further.
2. The problem statement is created using the 5-Why approach, driving as deep into the problem as data will permit.
3. An affinity or Ishikawa (fishbone) diagram is used to identify possible causes of the problem statement.
4. A problem description is written based on further investigation of the what, where, when and how big data collected (*is* and *is not*).
5. Possible causes on the affinity or Ishikawa (fishbone) diagram can then be reduced by using data from the problem description.
6. Theories are developed on remaining possible causes.
7. Root cause is verified by turning it on or off at will.
   a. PCAs are determined for root cause and inspection process (which also failed to stop the cause from escaping).
   b. Implementation and validation of the CA.

Preventive action

Often the root cause of a root cause is the system or lack of policies, practices, or procedures that supported the creation of the physical root cause. PA occurs after the physical root cause has been identified and PCA has been validated. PA recognizes the value of the information and actions taken during the CA function. This information is shared within the organization. Quality-One suggests the following steps for Preventive Action:

1. Capture the problem statement as an object–defect for searchable databases.
2. Link root causes to the problem statement with the PCA.
3. Identify other systems, facilities, and processes that could benefit from the knowledge captured.
4. Assure systems documents are updated, including but not limited to:
   a. failure mode and effects analysis (FMEA), and
   b. control plan methodology.

Work instructions

1. Archive information for future retrieval, including supporting information.
2. Publish and close out team experience.

chapter six

Key elements of APQP

Design FMEA (DFMEA)

What is it? DFMEA is a reliability tool that, among many functions, helps define, identify, prioritize and eliminate known or future failures of the system, subsystem or component. In essence it facilitates avoidance of failures.

Purpose: The purpose of the DFMEA is to perform a risk analysis of all reasonable design flows of the proposed product prior to manufacturing (Stamatis, 2015, p. 76).

When to use it: The primary use of DFMEA is to facilitate, with the appropriate and applicable team the following: (a) prevention planning, (b) changing requirements, (c) cost reduction, (d) increased throughput, (e) decreased waste, (f) decreased warranty costs (g) reduced non-value-added operations, and (h) avoidance of failures for future products from two perspectives: (1) failures perceived and (2) failures anticipated.

Process design

When dealing with process design, there are several key items that the analyst(s) must be concerned with. They are:

1. *Preparing the process*: This means that the team must be appropriately designated and the inputs of the process must be identified, such that:
   - *Team effort*: Manufacturing engineers, line operators, line supervisors, maintenance technician.
   - *Possible inputs to mapping*: Brainstorming; Operator manual(s); Engineering specifications; Operator experience; and Input from any of the 6Ms & S: Man, Machine (Equipment), Method (Procedures), Measurement, Materials, Mother Nature (Environment) and Safety considerations.

2. *PFC (Process Flow Chart)*: This must be done by walking through the process and not from memory or through what the process should be.
   - *What is it?* A visual diagram of the entire process from receiving through shipping, including outside processes and services.
   - *Purpose?* To help people "see" the real process. Process maps can be used to understand the following characteristics of a process: (a) Set-by-step process linkage; (b) Offline activities (measurement, inspection, handling); and (c) Rework, scrap.
   - *When to use it?* When you need to understand how a process is done. However, it must be at least initiated prior to completing the PFMEA.
3. *Process validation*: The flow chart should be able to identify where measurement analysis must be done, so be prepared for (a) MSA (Measurement System Analysis) and (b) MSA (Process Capability Study).

Process map summary

A process map provides understanding to inputs in:

- Potential Failure Mode Effect Analysis (PFMEA)
- Control Plan (CP)
- Capability Studies
- MSA Process Mapping helps us gain process knowledge!

Reviewer's checklist

- Process Flow must identify each step in the process. (They become the functions in PFMEA and control plan).
- Should include abnormal handling processes (scrap; rework; extended life testing).
- Process Flow must include all phases of the process (receiving of raw material; part manufacturing; offline inspections and checks; assembly; testing; shipping; and transportation).

Process FMEA (PFMEA)

What is it? A methodology (tool) used to identify and prioritize risk areas and their mitigation plans in the process.

Purpose: Identifies potential failure modes, causes, and effects. Inputs come from the DFMEA and process flow diagram. Identifies key inputs that positively or negatively affect the quality, reliability, and safety of a product or process. It also confirms the Special

Characteristics of Product/Process that impact the ultimate safety/performance of the end product.

When to use it? After completion of the process flow diagram and prior to tooling for production. Quite often the information from the DFMEA and or the PFC may not be available at the time of starting the PFMEA, which may cause some difficulties, but it should not be the reason for *not* doing a PFMEA. *Do* what you can with the info that you have.

Potential Failure Modes generally fall into six categories: (1) no function, (2) degradation over time (wear-out), (3) intermittent, (4) partial, (5) unexpected (surprise), and (6) over function. Therefore, it is very important to discuss with the team all credible Potential Failure Modes. The team should be able to pose and answer the following questions:

- How can the process/part fail to meet requirements?
  - Regardless of engineering specifications, what would a customer consider objectionable?
- In each instance, the assumption is made that the failure could occur, but will not necessarily occur. As such:
  - Each failure mode should be credible.
  - Do not list acts of God or freak accidents.
  - A description of non-conformance.
  - Assume incoming parts are correct.
  - Remember to consider subsequent operations—examples of failure modes include: potential failure modes should be described in "physical" or technical terms, not as a symptom noticeable by the customer. For example: burred, bent, hole off location, cracked, hole to shallow, hole missing, handling damage, dirty, hole to deep, surface too rough, corrosion, open circuit, and so on.

Potential effects are defined as the effects (consequences) of the failure on the customer(s), so:

- Describe them in terms of what the customer might notice or experience.
- State them clearly if the failure mode could impact safety or cause noncompliance to regulations.
- For the end user the effects should always be stated in terms of product or system performance such as for example: noise, rough, erratic operation, excessive effort, inoperative, unpleasant odor, unstable

operation, impaired, draft, intermittent operation, poor appearance, leaks, control impaired, rework, repairs, scrap, and so on.

- If the customer is the next operation, the effects should be stated in terms of process/operation performance, such as for example: cannot fasten, does not fit, cannot bore/tap, does not connect, cannot mount, does not match, cannot face causes, excessive tool wear, damages equipment, endangers operator, and so on.

Potential causes are defined as how the failure could occur and described in terms of something that can be corrected or controlled. Only specific errors should be listed, ambiguous phrases such as "operator error," "training," "machine malfunction," etc., should be avoided. Acceptable alternatives would be operator failed to install seal, or over temperature set incorrectly. The causes should be described so remedial efforts can be aimed at those causes that are pertinent. If the problem is operator dependent, then a strong recommendation would be to install a mistake-proofing device or sensor or something that will decrease the propensity of the error to surface.

Typical failure causes may include but are not limited to: improper torque (over/under), improper weld current, time, pressure, inaccurate gauging, improper heat treat (time, temperature), inadequate gating/venting, inadequate or no lubrication, part missing or miss located, worn locator, worn tool, chip on locator, broken tool, improper machine setup, improper programming, incorrect software version, or non-validated test system.

Analysis of PFMEA

The analysis of any FMEA is based on:

- Severity (of Effect, *not the cause*): severity of the effect on the customer and other stakeholders (higher value = higher severity). It is identified as "S." Severity—can only be improved by a design change to the product or process.
- Occurrence (of cause): frequency with which a given Cause occurs and creates Failure Mode (higher value = higher probability of occurrence). It is identified as "O." Occurrence—can only be reduced by a change that removes or controls a cause. Examples are redundancy, substituting a more reliable component or function, or mistake-proofing. In general, reducing the Occurrence is preferable to improving the Detection.
- Detection (capability of current controls): ability of the current control scheme to detect the cause before creating the failure mode and/

or the failure mode before suffering the effect (higher value = lower ability to detect). It is identified as "D." Detection—can be reduced by improving detection. Examples are mistake proofing, simplification, and statistically sound monitoring.

- Risk Priority Number (RPN): It is the product of S, O, and D. Even though it is not a very good measure of priority, unfortunately many organizations use it for that purpose. It may be used with graphical and statistical tools to show continual improvement.

The recommended classical evaluation of any FMEA is:

1. Severity.
2. Criticality (S × O).
3. RPN (S × O × D). DO NOT set a threshold for RPN: rather make sure you focus on continual improvement. DO NOT forget to address high Severity scores first.

Summary steps

1. For each Process Input, determine the ways in which the Process Step can go wrong (these are Failure Modes).
2. For each Failure Mode associated with the inputs, determine Effects on the outputs. Identify potential Causes of each Failure Mode.
3. List the Current Controls for each Cause.
4. Assign Severity, Occurrence, and Detection ratings after creating a rating key appropriate for your project. The classic rating scale may be found in the AIAG: FMEA (2008b) manual. The scale may be changed to reflect the organization's demand. However, if they are changed make sure you attach a copy of your ratings to the FMEA.
5. Calculate RPN.
6. Determine recommended actions to reduce high S, O, and RPNs.
7. Take appropriate actions and document.
8. Recalculate RPNs. Revisit steps 7 and 8 until all the significant RPNs have been addressed.

Reviewer's checklist

The reviewer's checklist may be used to:

- Verify there is a system for prioritizing risk of failure such as high S or high RPN numbers.

- Make sure that high S or RPN process concerns are carried over into the control plan Make sure that all critical failure modes are addressed.
    - Safety.
    - Form, fit, function.
    - Material concerns: see AIAG Core Tools for a detailed checklist.

Control plan

Since processes are expected to be updated as changes are made, control plans are *living* documents that need to be changed in step with the process—see Figure 6.1.

What is it? It is a tool used to define the operations, processes, material, equipment, methodologies, and special characteristics for controlling variation in key product or process characteristics within the manufacturing process.

Objective or purpose: It *communicates* the supplier's decisions during the entire manufacturing process from material receipt to final shipping; it *verifies* the existence of production controls at each step defined in the process flow/PFMEA; it *defines* reaction plans at each step should a nonconformance be detected; and it *denotes* special characteristics of the product/process that impact the ultimate safety/performance of the end product.

When to use it: After completion of the process flow diagram/PFMEA; at prototype, prelaunch and production; implementation of new process; and implementing a process change.

The interaction

Administrative: Identifies part number and description, supplier, required approval signatures, and dates.

- Phases: *Prototype*—a description of the dimensional measurements and material and performance tests that will occur during the prototype build.

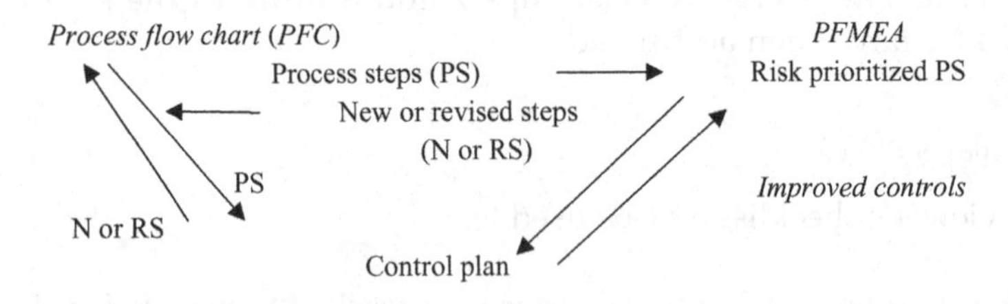

Figure 6.1 Interaction between PFC and PFMEA.

- *Pre-launch*—a description of the dimensional measurements and material and performance tests that will occur after prototype and before full production.
- *Production*—a comprehensive documentation of product/process characteristics, process controls, tests, and measurement systems that will occur during mass production.

Body of document: Since the control plan is linked to the flow chart and PFMEA, replication of the steps listed in those documents is done as the first step in producing your control plan. Therefore, each step listed in the PFMEA is documented in the same order on the control plan. In addition, any special characteristics listed on the PFMEA are replicated in the control plan as individual line items. For each step you determine the characteristics of either the product or the process or both that need to be controlled in order to repeatedly and reproducibly manufacture the component. If the feature has been denoted on the drawing or specification as a special characteristic by the customer or your internal analysis place the required symbol in the specification characteristic column.

- List the product specification tolerance required by the drawing or the process specification required to produce the product specification.
- List how you will measure or evaluate your product/process to determine if specification has been met.
- Specify the sample size and the frequency at which you will monitor the product produced at each step.
- List what documents the control. This could be a work instruction, a control chart, material certificate, set-up sheet, log sheet etc. Avoid statements such as TBD, operator training, operator error, unknown, or blanks.
- Provide specific guidance for the operator to carry out if a defect or issue is detected. Typical reaction plans include: segregate product, stop process, contact supervisor, scrap, contact engineering, rework, and no blanks.

Audit plans: Walk through the process. (*Gemba* [現場, also spelled *Genba*] is a Japanese term meaning "the actual place." Japanese detectives call the crime scene *Gemba*, and Japanese TV reporters may refer to themselves as reporting from *Gemba*. In business, *Gemba* refers to the place where value is created; in manufacturing the *Gemba* is the factory floor—a specific process.)

- Audit plans can be included in the control plan as a separate line.
- Auditing is an important tool for control.
- Process auditing should be a key element of the quality system of a business.

- Audits generally cover: effectiveness of controls; control plan (say) versus what is actually done (do).
- Audits should be objective (done by internal or external third parties if possible). Audit frequencies should be based on balancing level of risk (FMEA) and cost.

Reviewer's list

Remember the control plan is a planning tool. As such:

- Use it to decide what you should be doing.
- The AIAG format will help make sure the plan makes sense and is complete.
- Use the process flow diagram and PFMEA to build the control plan; keep them aligned.
- Controls should be effective. Keep it simple.
- Ensure that the control plan is in your document control system.
- Good control plans address all testing requirements—dimensional, material, and performance—and all product and process characteristics at every step throughout the process.
- The control method should be based on an effective analysis of the process, such as SPC, error proofing, inspection, sampling plan.
- Control plans should reference other documentation such as: specifications, tooling, procedures, instructions, etc.

Measurement system analysis (MSA)

- **What is it?** An MSA is a statistical methodology used to determine if a measurement system is capable of precise and accurate measurement. MSA is an analysis of the measurement process, not an analysis of the people!
- **Objective or purpose:** The objective is to determine how much error is in the measurement due to the measurement process itself. It quantifies the variability added by the measurement system. It can be used with both attribute and variable data.
- **When to use it:** On the critical inputs and outputs prior to collecting data for analysis; For any new or modified process in order to ensure the quality of the data. *Always remember* that MSA is an analysis of the measurement process, not an analysis of the people! *Important*!
- **Who should be involved:** Everyone that measures and makes decisions about these measurements should be involved in the MSA.

Special note: Do not confuse the measurement system analysis (MSA) with the Manufacturing Site Assessment (MSA). They are completely different!

In any MSA analysis there are two types of data:

1. Attribute Data Examples: They are binary in nature. Examples are: Count, Pass/fail, yes/no, red/green/yellow, timekeeping buckets. Generally, attribute data is not acceptable for PPAP submission.
2. Variable Data Examples: They are usually *quantities that are measured on a continuous and infinite scale.* Examples: physical measurement (length, width, area, etc.); physical conditions (temperature, pressure, etc.); physical properties (strength, load, strain, etc.); and continuous or non-ending.

The observed variation in process output measurements is not simply the variation in the process itself; it is the variation in the process plus the variation in measurement that results from an inadequate measurement system. Conducting an MSA reduces the likelihood of passing a bad part or rejecting a good part.

Observed variation: Differences between individual parts—often caused by: material variation, machine variation, set-up variation, or operator variation.

Precision:

- *Resolution*: The inability to detect small changes.
  - *Possible cause:* Wrong measurement device selected—divisions (resolution) on scale not fine enough to detect changes.
- *Repeatability*: The inability to get the same answer from repeated measurements made of the same item under absolutely identical conditions.
  - *Possible cause:* Lack of standard operating procedures (SOP), lack of training, measuring system variability. This variability is called equipment variation.
- *Reproducibility*: The inability to get the same answer from repeated measurements made under various conditions from different inspectors.
  - *Possible cause:* Lack of SOP, lack of training. This variability is called appraiser variation.
- Accuracy (location).
- Precision (variability).
- Linearity (a statistical method of modeling the relationship between a scalar dependent variable y and one or more explanatory variables (or independent variables) denoted X. It is used to find whether or not the data fit a straight line, square, or cube function).
- Bias stability.
- Process variation: Calibration addresses accuracy.

Gauge R&R

Gauge R&R is the combined estimate of measurement system repeatability and reproducibility (R&R). Typically, a three-person study is performed. Each person randomly measures ten marked parts per trial. Each person can perform up to three trials. There are two key indicators: (a) % P/T or measurement system or equipment variation, and (b) % R&R or process improvement or appraiser variation.

To actually gauge an R&R, there are forms that the AIAG provides in their published book of MSA. However, it is strongly recommended to use ANOVA. Make sure, however, that there are more than five distinct categories, otherwise your answers may not be correct.

Tips

Important: An MSA is an analysis of the process, not an analysis of the people. If an MSA fails, the process failed.

- A variable MSA provides more analysis capability than an attribute MSA. For this and other reasons, always use variable data if possible.
- The involvement of people is the key to success.
- Involve the people that actually work the process.
- Involve the supervision.
- Involve the suppliers and customers of the process.
- An MSA primarily addresses precision with limited accuracy information.

MSA checklist

- If the gauge/inspection measures a special characteristic or other important feature, then gauge the R&R.
- Make sure the study is recent—less than one year.
- Compare the control plan gages against the gauged R&Rs.
- % R&R and % P/T must be less than 30%.
- If you question that gauge, then question the technique and part sampling. Only as a last resort ask for additional studies.
- Measurement systems must be analyzed *before* embarking on process improvement activities.
- MSA helps understand how much observed variation is from the measurement system.
- MSA will tell you about repeatability, reproducibility, and discrimination. That is why it is important to have sampling during normal production to capture the total range of process variation.
- MSA assessors should be operators that would normally use the measurement system.
- MSA should be done on a regular basis.

Initial process study

Purpose

Capability studies can be used to define a problem or to verify permanent corrective actions in the problem-solving process. Specifically, they can be used to:

- Evaluate how well a process can produce product that meets specifications.
- Provide guidance about how to improve capability.
- Better process centering.
- Reduce variation.

Capability studies are measures of how well the process is meeting the design requirements. In performing a capability study, the team determines from sample data the process average and a spread (capability) of the process and compares this variation with the specifications.

Voice of the process (VP)

The normal distribution is the voice of the process—it's how the process behaves + the voice of the customer VC): the goal posts are the voice of the customer. They're the spec limits = Capability (VP *and* VC).

Capability studies

A short-term capability study covers a relative short period of time during which extraneous sources of variation have been excluded (guideline: 30–50; 0–50; 100–150 data points). A *long-term capability study* covers a longer period of time in which there is more chance for a process shift (guideline: 100–200 data points).

Capability versus performance

Capability ratios (C_P and C_{PK})

- Use a short-term estimate of sigma (s), $[\bar{R}/d_2]$, obtained from the within-subgroup variation.
- Show what the process would be capable of if it did not have shifts and drifts between subgroups.

Performance ratios (P_P and P_{PK})

- Use a long-term sigma (σ)—this is the actual std. deviation of the data obtained from within-subgroup plus between-subgroup variation.

Table 6.1 A comparison of critical versus noncritical criteria

	Critical	Noncritical	Decision
RED (bad)	<1.33	<1.00	Something is wrong, but it is not known what the problem is or how to fix it at this time. Needs attention *now.*
Yellow (marginal)	1.33–1.67	1.00–1.33	Something is not OK, but it is being attended to and there is a plan to fix it.
Green (good)	>1.67	>1.33	Everything is OK: proceed.

Note: For all critical items, the P_{pk} or C_{pk} must be higher or equal than 1.67. For noncritical items, the P_{pk} or C_{pk} must be higher or equal than 1.33.

- Show what the overall variation is—generally, performance ratios will be worse (smaller) than the corresponding capability ratios if the process has shifts and drifts. For a comparison of critical versus non-critical see Table 6.1.

Capability summary

- Capability ratios are used to compare the VC (specs) to the VP (natural process limits).
- For a capability ratio to be a good predictor of future performance, the process must be stable. Otherwise, the ratio is just a descriptor of past performance. It also must be normally distributed.
- The two key ways to improve process capability are to reduce variation and to improve centering.
- A capability ratio should never be interpreted without also looking at a control chart to verify stability and a histogram of the process to ensure normality. If you use MINITAB, the output of the capability is shown as well as the graphical interpretation. Do not overlook the graphs but more importantly the statistical (numerical) indices on the right lower corner of the screen. *Warning!* The distribution graph may look great, but the indices will tell you for sure if there is a normality function.
- The supplier should set warning tolerances and track changes to give a pre-emptive warning.

Initial process study: Checklist

- Ensure that the results are acceptable and that the process is stable and capable of producing a quality part.
- PPAPs should only be approved if the capability is greater or equal to 1.67 for critical dimensions and greater or equal to 1.33 for non-critical dimensions.
- Capability template is in the AIAG Core Tools.

PPAP (Production Part Approval Process)

PPAP is an automotive standard used to formally reduce risks prior to product or service release in a team-oriented manner using well-established tools and techniques. It was developed by AIAG in 1993 with input from the "Big 3"—Ford, Chrysler, and GM. The most current edition is the AIAG's 4th edition, effective June 1, 2006. The PPAP has been adopted by other industries with some modifications to reflect individual organizations.

Purpose

Provide evidence that all customer engineering design records and specification requirements are properly understood by the organization and to demonstrate that the manufacturing process has the potential to produce product that consistently meets all requirements during an actual production run at the quoted production rate.

When is PPAP required?

PPAP is required with any significant change to product or process, specifically, when

- New part
- Engineering change(s)
- Tooling: transfer, replacement, refurbishment, or additional
- Correction of discrepancy
- Tooling inactive > one year
- Change to optional construction or material
- Sub-supplier or material source change
- Change in part processing
- Parts produced at a new or additional location

Benefits of PPAP

- Helps maintain design integrity
- Identifies issues early for resolution
- Reduces warranty charges and prevents cost of poor quality
- Assists with managing supplier changes
- Prevents use of unapproved and nonconforming parts
- Identifies suppliers that need more development
- Improves the overall quality of the product and customer satisfaction

Full PPAP

A full PPAP covers 18 items and they are found in the publication AIAG: PPAP 4th ed.:

1. Design Records
2. Engineering Change Documents
3. Engineering Approval
4. Design FMEA
5. Process Flow Diagrams
6. Process FMEA
7. Control Plan
8. Measurement System Analysis Studies
9. Dimensional Results
10. Records of Material/Performance Test Results
11. Initial Process Studies
12. Qualified Laboratory Documentation
13. Appearance Approval Report
14. Sample Production Parts
15. Master Sample
16. Checking Aids
17. Customer-Specific Requirements
18. Production Part Approval Process Submission Warrant (PSW)

Submission levels

According to the AIAG: PPAP (p. 17) there are five levels:

- *Level 1*: Warrant only and appearance approval report as requested. Applied to: noncritical parts, noncritical raw/bulk material or catalog/commodity parts for electrical applications and recertification of existing parts previously approved at levels 3, 4, or 5. This level is reserved only for self-certified suppliers.
- *Level 2*: Warrant with product samples and limited supporting data. Applied to: critical bulk products such as paint/resin/chemicals, critical fasteners, simple material changes, simple revision level only changes or simple print updates not impacting form-fit-function. This level can also be applied to low risk parts within a product family.
- *Level 3*: Default submission level: warrant with product samples and complete supporting data. Applied to: new parts, changes affecting form-fit-function, reliability or performance. All products resourced to new suppliers, serial production parts, and existing high-risk parts undergoing a part number change.

- *Level 4*: Warrant and other requirements as specified by Customer Specific Requirements. This level is reserved for special applications only. Applied to: this level can only be applied with prior approval from supplier quality management.
- *Level 5*: Warrant with product samples and complete supporting documentation reviewed at the supplier's manufacturing location. On-Site.

Risk

High risk

- Parts associated with multiple critical features, complex design, or high-end technology that is not yet established in the general manufacturing environment
- Supplier's quality system and/or quality performance is not to Eaton satisfaction
- Critical process being conducted, e.g., heat treatment, leak proof welding
- Parts that impact the safety performance of the final product

Medium risk

- Parts that have at least one critical feature
- Parts that impact functional performance of the final product

Low risk

- Parts that have no critical features and can be manufactured by any manufacturer in the commodity category
- Catalogue parts
- Supplier's quality system is acceptable
- Supplier's quality performance can be demonstrated over time

PPAP status

Approved

- The part meets all customer requirements.
- Supplier is authorized to ship production quantities of the part Interim.

Approval permits

- Shipment of part on a limited time (90 days) or piece quantity basis. Submission must have a specification deviation identifying permanent corrective action to achieve full approval within 90-day period.

Rejected

- The part does not meet customer requirements based on the production lot from which it was taken and/or accompanying documentation.

Functional testing at the plant level must be approved before a supplier begins shipping product to the assembly. It is possible to have a signed PSW but not approved functional testing.

Quality planning and product approval

General requirements

- Suppliers *must* use APQP.
- Suppliers *must* approve parts through PPAP.
- Suppliers *must* retain records Life of Product.
- Suppliers *must* notify and obtain approval prior to implementing changes.

Supplier assessment and qualification

The supplier assessment and qualification process includes:

- Initial supplier profile—accessed through the customer's own system.
- Supplier screening/data analysis process: (a) supplier's current delivery performance based on 100% on time delivery (OTD) expectation; (b) supplier's quality performance for previous 12–24 months; (c) supplier's registration to an industry sector quality system; (d) cost competitiveness; and (e) supplier's financial strength for future growth.
- Supplier assessment: Typically consists of an on-site audit (OSA).
- Assessment results/timely corrective actions.
- Approvals: (a) full approval (b) conditional approval, and (c) un-approved (approval can be lost to those previously approved).

Cost of quality

The following is a list of potential cost of poor quality charges (*not* exhaustive!):

- Inspection
- Sorting
- Rework
- Line disruption
- Premium freight
- Third-party inspection (cost of increased inspection)
- Premium product cost paid to support production
- Downtime/overtime
- Unexpected downtime
- Equipment breakage
- Travel warranty costs
- Containment activities

chapter seven

PPAP with the specific requirements of GM, FCC, and Ford

The PPAP is a summary of the APQP process, and it follows the flow of AIAG: PPAP 2009, 4th ed. It is a document with high-level information that the customer uses to validate the design and process of a given supplier. It is generally in the form of a booklet (hard copy, however it may be electronically transmitted) and it typically covers the following sections. (Of course, depending on the organization, customer and product there may be more or less content than those identified here).

I. General Content
   - I.1. Purpose, Applicability and Approach
   - I.2. PPAP Process Requirements
   - I.3. Customer Notification and Submission Requirements
   - I.4. Submission to Customer—Levels of Evidence
   - I.5. Part Submission Status
   - I.6. Record Retention

II. Customer Specific Requirements (as required)
   - II.1. Fiat Chrysler Corporation-Specific Instructions
   - II.2. Ford–Specific Instructions
   - II.3. General Motors-Specific Instructions
   - II.4. Truck OEM-Specific Instructions

III. Appendices
   - Appendix A: Completion of the Part Submission Warrant
   - Appendix B: Completion of Appearance Approval Report
   - Appendix C: Dimensional Report
   - Appendix D: Material Test Report

Appendix E: Performance Test Report
Appendix F: Bulk Material-Specific Requirements
Appendix G: Tire Industry-Specific Requirements

Glossary (as necessary)
Specific or special acronym list used in the organization's (as required) PPAP

Purpose, applicability and approach

Purpose

Definition of generic requirements for production part approval to determine:

1. Proper understanding of specification requirements and design records by the supplier.
2. That the supplier's process has the potential to produce product that consistently meets requirements during actual production run at the quoted production rate.

Applicability and approach

PPAP shall apply to internal and external supplier sites of bulk materials, production material, production, or service parts

1. Standard catalogue production shall comply with PPAP unless formally waived by the customer.
2. Tooling shall be maintained for standard catalogue items as long as the items are offered or stated as being available.

All questions about PPAP should be addressed to the customer product approval activity or the supplier technical estimate (STA).

When PPAP is required

The supplier shall obtain full approval from the customer product approval activity for:

1. A new part or product (i.e., a specific part, material, or color not previously supplied to the specific customer).
2. Correction of a discrepancy on a previously submitted part.

3. Product modified by an engineering change to design records, specifications, or materials.
4. Any situation required by Section I.3 (Customer Notification and Submission Requirements, pp. 11–14).

PPAP process requirements

1. Production parts are manufactured at the production site using the production tooling, gauging, process, materials, operators, environment, and process settings (e.g., feeds/speeds/cycle times/pressures/temperatures).
2. Parts for production part approval shall be taken from a significant production run. This run shall be taken from one hour to one shift's production, with the specific production quantity to total a minimum of 300 parts unless otherwise specified by the authorized customer quality representative.
3. Parts from each unique production process, e.g., duplicate equipment, each position of a multiple cavity die, mold, tool, or pattern shall be measured, and representative parts tested.

Significant production run

A significant production run must be taken from the actual production. This can be from one hour of production, or 300 parts consecutive or as per the authorized customer representative—STA or SQE (AIAG: PPAP 2009 p. 3). A PSW run is conducted at the intended point of production using the production machinery, tools, and facilities with the intended people at the intended production rate that is capable of meeting the customer's daily production rate (as specified on the request for quotation) and results in components that are correct to drawing and meet all customer's specifications. Typically, the volume produced will be sufficient to supply customer with one full day's production.

Evaluation of first time through (FTT), total actual cycle time (TAKT), total estimated cycle time (TACT). The formula for TACT is the same as that of the TAKT; however, the data for the calculation more often than not are based on surrogate data. A third evaluation is the OEE to check that supplier's capacity can be supported by current machine availability (Stamatis, 2010).

TACT time run

A TACT time run (run @ rate) is conducted at the intended point of production using the production machinery or surrogate, with minimum of one tool (out of several planned for the same process due to capacity) and facilities with the intended people at the intended production rate that is capable of meeting the customer's TAKT (volume produced—generally 300 parts).

Evaluation of FTT, TAKT, and OEE to check supplier's capacity can be supported by STA.

The PSW

The four main PSW items (production parts from the TACT time run and the significant production run will be taken for the following items:

1. Statistical process package: *Required*—all CP items (e.g., CCs, SCs) or other values/checks; *not recommended*—attributive tests, in fact, they should be avoided; *strongly recommended*—if there are issues, concerns, or even problems with capability and or capacity always address them with your customer representative. remember that without proof of capacity and capability the PSW *will not* be signed.
2. Dimensional measurements: *All* dimensions on parts and product materials with dimensional requirements must be accounted for.
3. Appearance approval: Final approval by styling quality (craftsmanship quality, only for "appearance items").
4. Material and functional tests: Engineering specification (ES) tests with single results—no attributive tests! Smell test for interior parts.

The PPAP process requirements

The following documents and items must be completed by the supplier for each part

1. All following 18 items must be completed and be readily available for customer use. For some customers the APQP, PPQI (prototype process potential & quality indexes) sheets are also required.
2. Any results that are outside specification are cause for the supplier not to submit the parts, documentation, and/or records.
3. If the supplier is unable to meet any of these requirements, the customer shall be contacted for determination of appropriate corrective action.

4. Inspection and testing for PPAP shall be performed by a qualified, accredited laboratory, and/or facility and the approval of the customer representative.
5. The laboratory test results shall be submitted in the laboratories format; they may be externally or internally tested.
6. Blanket statements of conformance are unacceptable for any test results.

1. *Design records*
    a. The supplier shall have all design records for the saleable product, including design records for components or details.
    b. For all data in electronic format the supplier shall produce a hard copy to identify measurements taken.
2. *Engineering change documents*: The supplier shall have any authorized engineering change documents not yet recorded in the design record but incorporated in the product, part, or tooling. There are two possibilities here. The change may be initiated by the PD, in which case it is called an *alert* (generally it is good for 90 days; if nothing happens during that time, it may be renewed or cancelled), or it can be initiated by the supplier, in which case it is called the customer engineering approval process.
3. *Customer engineering approval*: Where specified by the design record, the supplier shall have evidence of customer engineering approval.
4. *Design FMEA (Failure Mode and Effects Analysis)*
    a. The supplier shall have a DFMEA for parts or materials for which they are design-responsible.
    b. The DFMEA shall be developed in accordance with the AIAG or the VDA.6 requirements. In the case of the Ford Motor Company, the supplier should be aware of Ford's requirements as well.
5. *Process flow diagrams*
    a. The supplier shall have a process flow diagram in supplier-specified format that clearly describes the production steps and sequence, as appropriate, and meets the specified customer needs, requirements, and expectations.
    b. Process flow diagrams for families of similar parts are acceptable if the new parts have been reviewed for commonality.
6. *Process FMEA*
    a. The PFMEA shall be developed in accordance with the AIAG or the VDA.6 requirements. In the case of Ford Motor Company, the supplier should be aware of Ford's requirements as well.
    b. A single process FMEA may be applied to a process manufacturing a family of similar parts or materials.

7. *Dimensional results*
   a. All dimensional verification required by the Design Record and the CP shall be completed and results indicate compliance with specified requirements.
   b. The supplier shall have dimensional results for each unique manufacturing process (including multi-tooling).
   c. Indications required:
      i. Date of the design record
      ii. Change level, authorized engineering change documents
      iii. One of the parts measured to be identified as the master sample
   d. Supplemental documents (if needed) shall show:
      i. Change level
      ii. Drawing date
      iii. Supplier name and part number
   e. For dimensional results use the form in Appendix C (AIAG: PPAP, p. 29) or a checked print.

8.1. *Material test results*: Tests shall be performed for all parts and product materials when so required by the design record or CP. Material test report shall indicate:
   a. Design record change level of the parts tested
   b. Number, date and change level of the specifications
   c. Date of testing, material subcontractor's name

   On the other hand, for products with a customer-approved source list, suppliers shall procure materials/services from the subcontractor on that list. For the form for material test results see AIAG: PPAP, Appendix D, p. 31.

8.2. *Performance test results*

   Performance tests shall be performed for all parts and product materials when so required by the design record or CP. The test report shall indicate:
   a. Design record change level of the parts tested
   b. Number, date and change level of engineering specification
   c. Date on which the testing took place

   Results should be listed on an understandable format and include the quantity tested (see AIAG: PPAP, Appendix E, p. 33).

8.2.1. *Bulk material requirements checklist*
   a. For bulk material, the checklist in AIAG: PPAP, Appendix F, shall be jointly agreed upon by the customer and supplier.
   b. All specified requirements shall be completed unless specifically indicated as "not required" (NR) on the checklist.
   c. Additional requirements may be specified on the checklist.

9. *Initial process studies*
   a. The level of initial process capability or performance shall be determined to be acceptable prior to submission for all special characteristics designated by the customer or supplier.
   b. The supplier shall perform measurement system analysis to understand how measurement error is affecting the study measurements.
   c. See AIAG: PPAP, Notes 1–5, p. 7.

9a. *Initial process studies—quality indices*
   a. C_{pk} is the capability index of a stable process—short term.
   b. P_{pk} is the performance index—long term.
   c. P_{pk} is the preferred index for it accounts for the true standard deviation as opposed to the C_{pk}, which uses the approximate value of ($\bar{R}/d_2$) and can only be used for stable processes!
   d. Only if the data are normally distributed can C_{pk} and P_{pk} be calculated.
   e. Statistical process control (see Stamatis, 2002; AIAG: SPC reference Manual or any other SPC reference).

9b. *Initial process studies—acceptance criteria*
   The following acceptance criteria shall be used for evaluating initial process study:

Index Value >1.67	Meets customer requirements.
1.33 ≤ (Index Value) ≤1.67	Currently acceptable, but may require some improvement.
Index Value <1.33	Does not meet the acceptance criteria. Contact your customer representative!

9c. *Initial process studies—unstable processes*
   a. Unstable processes may not meet customer requirements.
   b. The supplier shall identify, evaluate, and eliminate special causes of variation prior to PPAP submission.
   c. The supplier shall notify the customer of any unstable process and shall submit a corrective action plan to the customer prior to any submission.
   d. Warning! In some cases, even though the process may be out of control, it may still be within specification. That is a very dangerous assumption to make and can create many problems downstream. The process is *not* stable, therefore it may be moving in all directions and it just so happens that the data in question may be randomly acceptable. The correct action is to bring the process back to stability and then proceed.

9d. *Initial process studies—processes with one-side specifications or non-normal distributions*
The supplier shall determine with the customer an alternative acceptance criterion. Unless the supplier has a qualified statistician or a Six-Sigma black belt *do not* try to transform the data. That is also a dangerous step.

9e. *Initial process studies—strategy when acceptance criteria are not satisfied*

a. The supplier shall contact the customer if the process cannot be improved.
b. If acceptance criteria cannot be attained, a corrective action and control plan shall be submitted to the customer (providing a 100% inspection). Without it there will be *no full PSW!*
c. Continuous improvement for additional techniques or possibly changing the process.
d. Continue variation reduction efforts until a C_{pk}/P_{pk} of 1.33 or greater is achieved, or until customer's full approval.

10. *Measurement system analysis studies*
The supplier shall have applicable MSA studies (e.g., gauge R&R studies) for all equipment used for new or modified gauges, measurement, and test equipment (see AIAG: MSA reference Manual). The preferred method is to use the ANOVA approach.

11. *Qualified laboratory documentation*
The supplier shall have a laboratory scope and documentation showing that the laboratories used comply with all appropriate standards, including the latest version of ISO/IEC 17025. Some suppliers may not require it: check with your customer representative.

12. *Control plan*
The supplier shall have a CP that defines all controls used for process control and complies with international as well as customer's requirements. For some customers:
    - CP approval is required in any case prior to the production part submission date.
    - Component review team (CRT) agreement on SC/CC: engineering, manufacturing, customer representative and supplier. The linkages are shown in Table 7.1.

13. *Part submission warrant* (*PSW*)
    a. Upon satisfactory completion of all required measurements and tests, the supplier shall record the required information on the PSW form.
    b. A separate PSW shall be completed for each part number unless otherwise agreed to by the customer.

Table 7.1 The linkages of the control plan

DFMEA →	PFMEA →	Production control plan	Customer needs ←
		• Avoid attributive tests! (yes/no tests) • Should be a physical dimension, with clear tolerances • Must include all critical and significant characteristics • Action plan for all "scrap parts" (zero defect planning)	Customer manufacturing end user requirements Component review team (CRT)

c. The supplier shall verify that all of the measurements and test results show conformance with customer requirements.
d. A responsible supplier official shall approve the PSW and provide date, title, and telephone number.
e. PSW is the verification that all PPAP documents are available.

If production parts will be produced on more than one die, line, cavity, mold, pattern or production process, one part from each shall be evaluated as such identified and documented on the PSW or attachment.

13a. *Part submission warrant (PSW)—part weight*
a. The part weight in kg (0.0000) shall be recorded unless otherwise specified by the customer.
b. The weight is the net weight of the part only.
c. To report the average weight, a sample size of ten parts shall be measured.
d. At least one part shall be measured from each cavity, tool, line or process to be used in product realization.
e. PSW form; see Appendix A.

14. *Appearance approval report (AAR)*
a. Typically applies only for parts with color, grain, or surface appearance requirements.
b. A separate AAR shall be completed if the part has been designated by the customer as an "appearance item."
c. The completed AAR and representative production parts are to be submitted to the location specified by the customer.
d. Additional requirements may be recorded in customer-specific requirements.
e. AAR form; see Appendix B.

15. *Sample production parts*
    The supplier shall provide sample products as requested by the customer and as defined by the submission request.
16. *Master sample*
    A master sample shall be retained for the same period as the production part approval records:
    a. Until a new master sample is produced for the same part number for customer approval.
    b. Where a master sample is required by the design record, CP, or inspection criteria, as a reference or standard to be used.
    c. The master sample shall be identified as such and show the customer approval date on the sample.
    d. The supplier shall retain a master sample for each position of a multiple cavity die, mold, and tool or production process.
17. *Checking aids*
    a. If requested, the supplier shall submit with the PPAP submission any part-specific assembly or component checking aids.
    b. All aspects of the checking aids shall be agreed with part dimensional requirements, as well as any engineering changes.
    c. Preventive maintenance of checking aids shall be provided.
    d. Measurement systems analysis studies shall be conducted in compliance with customer requirements (MSA manual).
18. *Customer specific requirements* (*CSRs*)
    The supplier shall have records of compliance to all applicable CSRs, if available and required.

Submission to customer

1. The submission level prescribes what the supplier must submit to the customer. Self-certifying suppliers are at the manufacturing site. This level is designated Level 1. Level 3 is the default, and it means that some items will be self-certified, where some are going to be very thoroughly reviewed by the customer representative. Levels 2–5 are generally for suppliers with historical problems, either because they are new suppliers or the products they produce have special requirements, are high impact items or are part of a major vehicle launch. *All* Levels 2–5 suppliers need STA sign off. If there are tests involved, then a PD has to sign off as well. Remember that STA or the supplier quality engineer (SQE) is responsible for PSW approval.

2. The customer will identify the submission level that will be used with each supplier. That is done through the PD. Most customers have their own supplier improvement metrics and may be found in the intranet site through the Covisint supplier portal.
3. The customer's choice of levels for a supplier will be determined by such factors as:
   a. Supplier historical compliance with all requirements
   b. Supplier quality recognition status (e.g., Q1 3rd ed.)
   c. Part criticality (e.g., high impact)
   d. Experience with prior part submissions
   e. Supplier expertise with the specific commodity
   f. Whether or not the supplier has experience with the customer (new or old supplier)
   g. It is possible that different customers will assign different submission levels to the same supplier manufacturing location

Part submission status

Suppliers shall be notified by the customer as to the disposition of the submission. After production part approval, suppliers shall assure that future production continues to meet all customer requirements.

For those suppliers that have been classified as "self-certifying" by a specific customer, submission of the required documentation (PPAP Level 1) showing supplier approval will be considered as customer approval unless the supplier is advised otherwise. Just because the supplier is Level 1 does not mean that they may skirt some of the requirements. It is important for the supplier that all requirements must be completed. This means that for all intents and purposes Level 1 and Level 5 have exactly the same requirements. The difference is that the customer representative does not sign the PSW for Level 1.

Full approval

Production approval indicates that the part meets all customer specifications and requirements. However, for delivery of the part there is one more hurdle. That is, functional testing must be completed at the customer's plant.

The supplier is therefore authorized to ship production quantities of the part subject to releases from the customer scheduling activity. This is profoundly important, because unless the functional testing at the customer's facility is complete, the supplier cannot provide that part for production.

Always add the OEE calculation sheet!

Interim approval

Interim approval permits shipment of material for production requirements on a limited time or piece quantity basis. Interim approval will only be granted:

- When the root cause is clearly defined
- An interim action plan is agreed by the customer
- Resubmission to obtain "production approval" is required.
- Interim PSW (capacity or design/quality—this process starts at 1PP)
- Some kind of alert is necessary, describing *all* deviations and risks (design, testing, process) from final PSW—most customers have their own system for this alert and it is computerized
- Alert is issued and signed off by engineering/launch Team, with the sign off that engineering accepts the parts for our vehicle production (time limited!)
- Supplier provides facility; tool, operator changes and risk assessment with a PSW timing chart
- Team (supplier, engineering, and customer representative) decides if part is saleable/not saleable
- Supplier adds alert number to interim PSW documentation
- Customer representative sign off for high impact parts
- OEE calculation sheet is always required

Rejected

1. Means that the submission/documentation does not meet customer requirements
2. Corrected product and documentation shall be submitted and approved before production quantities may be shipped

PSW verification

Recommended PSW verification matrix is shown with a Tick (check mark) in a YES/NO box on PSW form for submission results:

Full PSW	YES
Interim (capacity)	YES
Interim (design or quality)	NO
Rejected	NO

Record retention

1. Production part approval records (all 18 requirements), regardless of submission level, shall be maintained for the length of time that the part is active plus one calendar year. This may also depend on the specific customer requirement.
2. The supplier shall ensure that the appropriate PPAP records from a superseded part PPAP file are included or referenced in the new part PPAP file.

Specific customer requirements

As we have mention earlier, the international or industry requirements may not cover each customer's needs. As a consequence, most, if not all customers, have their own. Therefore, let us look at some of Ford's, FCA's, and GM's key requirements, that their suppliers are expected to follow.

Ford CSRs for IATF 16949

The core Ford requirements are in the following categories:

Substance use restrictions

A statement indicating conformance with Ford Engineering Material Specification WSS-M99P9999-A1 is required on the material test report.

Plastic parts marking

1. Suppliers of plastic parts are encouraged to mark their parts with the appropriate ISO symbols
2. Further information is provided in material specification E-4, available online at: https://web.keyinfo.ford.com/manuals/index.html
3. Affected suppliers are requested to indicate compliance on the PSW form, as appropriate

PSW for smell-related items

1. All sealing and interior parts out of production processes shall be reviewed/tested in view of smell.
2. Odor self-assessment to be performed according to system design specification.
3. FLTM BO 131-01.
4. All parts having smell criteria should be reviewed by the Ford Central Lab.
5. For all of those smell-related items, the supplier shall contact the customer representative to follow local procedure (e.g., tests incorporated into CP).

CPs

1. Shall be developed by the supplier and be available for review by the customer as early as possible, and in any case prior to the production part submission date.
2. Control plans must include all critical and significant characteristics and have to be signed by product engineering.
3. Some Ford quality activities may request that a copy of the signed CP be attached to the PSW and forward to the customer part approval activity.
4. [∇/CC (inverted Delta) used for Critical Characteristics; SC used for Significant Characteristics].

Appearance item approvals

1. All parts having appearance criteria (e.g., visible for consumer) shall be reviewed by the Ford Design Quality Office.
2. The completed Appearance Approval Report shall accompany this submission.
3. After approval signatures have been obtained from the designated Ford Design Quality representative, the form shall be included with Warrant.

Part submission level

1. It is important that suppliers designated as Level 2–5 account for any additional time that may be required to obtain customer representative approval, when providing the PSW promise date.
2. If the customer representative engineer is unavailable to approve the PPAP, the supplier should contact the customer representative commodity manager who may assign another engineer.
3. Internal operations should contact their responsible approval activity or the quality office regarding PPAP submission level.

Manage the change

1. The supplier shall develop and implement a process to manage change to ensure that any product affected by any revision in design or process shall continue to meet all specifications.
2. All design changes shall be clearly described and reference the Worldwide Engineering Release System (WERS) number under the "Additional Engineering Changes" on the PSW form.
3. The customer using the facility may request full or partial PPAP documentation from the supplier—supplier data shall be readily available upon request.
4. All post job #1 changes (e.g., running changes) shall obtain functional approval from the customer using facility prior to shipping of production quantities.

5. At vehicle operation facilities contact the plant vehicle team (PVT) for functional approval.
6. If the component is used at multiple Ford facilities, all using facilities must provide functional approval.
7. Functional approval is provided at the bottom of the PSW form under "For Customer Use Only."
8. See supplier checklist/approval for managing change as prescribed in the Ford Motor Company - Customer-Specific Requirements - For IATF-16949:2016 (effective May 1, 2017 page 30).

Family of parts

Suppliers are permitted to submit multiple part numbers (same family of parts) on a single PSW with all parts specifics (e.g., prefix, base, suffix) clearly noted on the PSW.

Labeling requirement

1. Suppliers to Ford European facilities are required to affix orange labels (Form EU 3441, minimum A5-size) on all four sides of the packaging for all shipment of new or changed product to each using Ford facility.
2. Powertrain suppliers are required to include their unique supplier generated PSW number on each label of the PSW shipment (ongoing shipments are excluded from this requirement).

Supplier request for engineering approval (SREA)

1. The SREA procedure applies to all internal and external suppliers without online WERS capability.
2. Ford product engineering approval of a SREA is required prior to implementing the change.
3. Once Ford product engineering determines that the change is feasible and an SREA is required, the supplier will complete and submit the SREA, Form 1638, to the responsible design engineer.

SREA/process changes

1. The supplier is empowered to implement process changes without issuing and obtaining SREA approval with the following exceptions:
   a. Changes in heat treat, plating/coating, and solderability
   b. Changes to the manufacture of supplier-designed electronic components

2. Relocation of product to a different manufacturing location. These exceptions will be waived for full service suppliers
3. Control items or emission components always require SREA

These general categories are identified in the following elements of the standard. They are:

ELEMENT 4: CONTEXT OF THE ORGANIZATION

4.1–4.2 NO CHANGES TO THE REQUIREMENTS

4.3 DETERMINING THE SCOPE OF THE QMS

The structure of this document aligns with the requirements in the applicable sections of IATF 16949. Several section headers are followed by the statement "No Ford Customer-Specific Requirement for this section" to verify that there is no auditable Ford-specific requirement for the section. The presence of this statement does not mean that no other commercial or technical requirements exist for the subject addressed in the section, or that this statement supersedes existing commercial or technical requirements. Tooling and equipment suppliers to Ford Motor Company are not eligible for certification to IATF 16949. Registration to ISO 9001 is acceptable.

Third-Party Registration: To achieve Q1 3rd ed. (refer to https://web.qpr.ford.com/sta/Q1.html), production and service part organizations supplying product to Ford shall be third-party registered to IATF16949 through an IATF-recognized certification body. The official list of IATF-recognized certification bodies is available through http://www.iatfglobaloversight.org/certBodies.aspx. A sub-tier supplier hired by the organization to perform services not directly related to a Ford Motor Company contract (e.g., floor cleaning or grass cutting) is not impacted in any way by the sub-tier supplier development or other sub-tier supplier requirements stated in IATF 16949. Evidence of IATF 16949 certification verification organizations shall record evidence of their certification to IATF 16949 in GSDB Online, available through Ford Supplier Portal: https://web.gsdb2.ford.com/GSDBeans/servlet/gsdbeans.web.lib. GSDB. Notification of IATF 16949 registration status change organizations shall notify Ford of any change in their IATF 16949 registration status by updating their certification information in GSDB Online. Such changes include, but are not limited to:

- Initial certification
- Recertification
- Transfer of certification to a new certification body

- Certificate withdrawal
- Certificate cancellation without replacement

IATF 16949 Certification Waiver: Ford may, at its discretion, fully waive certain organizations from IATF 16949 certification. This waiver generally applies to those organizations whose QMS is acceptable without certification to IATF 16949. However, Ford still requires suppliers to be certified—eventually. Identification of candidate organizations for waiver from IATF 16949 certification is the responsibility of Ford. Verification and maintenance of waiver status is the responsibility of Ford.

4.3.1–5.1.1 No changes to the requirements

5.1.1.1 Corporate responsibility

The organization shall comply with basic working conditions in the global terms and conditions and the related supplier social responsibility and anti-corruption requirements.

1.1.1.2–5.1.1.3 No changes to the requirements

5.1.2 Customer focus

The organization shall demonstrate enhanced customer satisfaction by meeting the continuous improvement requirements of Q1 3rd ed.—minimum score is 80/100 points, as demonstrated in the organization's QOS.

5.2–5.3 NO CHANGES TO THE REQUIREMENTS

5.3.1 Organizational roles, responsibilities and authorities—supplemental

The organization shall notify Ford Motor Company Supplier Technical Assistance in writing within ten working days of any changes to senior management responsible for product quality or company ownership.

5.3.2–6.1.2.2 No changes to the requirements

6.1.2.3 Contingency plans

The organization shall notify the Ford receiving plant, the buyer, and the STA engineer within 24 hours of organization production interruption. The organization shall communicate the nature of the problem to Ford and take immediate actions to assure supply of product to Ford.

> **Note:** Production interruption is defined as an inability to meet the Ford specified production capacity volume.

The supplier shall have a documented supply risk management operating system in place and it should include:

- The application of the requirements for risk analysis, preventive actions and contingency planning described in Sections 6.1.2.1 through 6.1.2.3 of IATF 16949:2016 through the organization's supply chain.
- Documentation of the organization's supply chain (supplier name, location, and parts) for all Ford-specified parts and associated raw materials.
- A system to assess and monitor supply chain financial and operational risks.
- A list of Ford's endorsed supply chain monitoring services and tools is available through Ford's website. At all times, Ford reserves the right to review the documented information of the supply chain risk assessment reviews.

6.2–7.1.3 No changes to the requirements

7.1.3.1 Plant, facility, and equipment planning

Capacity reporting: whenever the organization reports purchased part capacity (average purchased part capacity—APPC, or maximum purchased part capacity—MPPC) to Ford in demonstration of compliance to the APW/MPW capacity requirements, the organization shall use the capacity analysis report to determine the values of APPC and MPPC reported. Where equipment is not dedicated to the Ford part being reported for PPC, the organization shall use either the shared loading plan in the capacity analysis report or the detailed shared loading tool. The report may include:

- Quarterly reporting of PPC to Ford's capacity planning systems
- Responding to a request for quote
- Responding to a capacity study
- Capacity verification associated with PPAP
- Any other Ford request for reporting purchased part capacity

Note: For the APPC and MPPC to be acceptable, they must meet or exceed the required capacity—APW in a five-day operating pattern and MPW in six-day operating pattern respectively. Whoever is responsible for completing the report is required to have completed the latest CAR training. Planners are required to review the CAR training annually.

7.1.4–7.1.5.1 No changes to the requirements

7.1.5.1.1 Measurement system analysis

Gauging requirements: All gauges used for checking Ford components/parts per the CP shall have a gauge R&R performed in accordance with the appropriate methods described by the latest AIAG Measurement Systems Analysis Manual (MSA) to determine measurement system variability. The gauge R&R is to be completed using Ford-produced parts. The CP identifies which gauges are used for each measurement. Any measurement equipment not meeting the MSA guidelines must be approved by STA.

Family of gauges: where multiple gauges of the same make, model, size, method of use and application (including range of use) are implemented for the same part, use of a single gauge R&R covering those multiple gauges (family of gauges) requires STA approval.

Parts and operators for gauge R&R studies: Minimum requirements are:

1. Variable gauge studies should utilize a minimum of 10 parts, 3 operators and 3 trials.
2. Attribute gauge studies should utilize a minimum of 50 parts, 3 operators, 3 trials.

7.1.5.2–7.1.5.3.1 No changes to the requirements

7.1.5.3.2 External laboratory

The organization shall approve commercial/independent laboratory facilities prior to use. The acceptance criteria should be based on the latest ISO/IEC 17025 (available through ISO, http://www.iso.org/), or national equivalent, and shall be documented. Accreditation to ISO/IEC 17025 or national equivalent is not required.

7.1.6–7.2 No changes required

7.2.1 Competence—supplemental

Training shall include the appropriate Ford systems.

7.2.2–7.5.1.1 No changes required

7.5.2 Creating and updating

Note 1: Where the organization uses Ford documents/instructions or other documents of external origin, the organization ensures that the appropriate revision level is used—either the most current version available from the FSP (Ford Supplier Portal, https://fsp.covisint.com) or as specified by the Ford Motor Company.

Note 2: Engineering Standards may be obtained from the following sources:

- IHS Markit (http://www.ihs.com/)
- ILI Infodisk Inc. (http://www.ili-info.com/)

If any standards are not available through the above sources, organizations should contact Ford engineering, or for organizations with Ford Intranet access, http://www.rlis.ford.com/cgi-bin/standards/iliaccess.pl/ may provide a more complete inventory. Ford Engineering Specifications may be available in Ford's CAD database, TeamCenter; contact the Ford PD engineer for details. Additionally, Ford Engineering CAD and Drafting Standards (FECDS) are available through https://team.extsp.ford.com/sites/C3PNGMethods/C3PNGMethods.html.

Engineering specifications

Ford requires all manufacturing sites to report all materials per WSS-M99P9999-A1, as noted in PPAP, Ford Specific Instructions. These requirements are detailed on Ford Supplier Portal, https://fsp.covisint.com (Important Documents—RSMS Communication Package).

Engineering specification (ES) test performance requirements

The goal of ES testing is to confirm that the parts meet design intent. ES test failure shall be cause for the organization to stop production shipments immediately and take containment actions. The organization shall immediately notify Ford Engineering, STA, and the Ford Motor Company facility of any test failure, suspension of shipments, and identification of any suspect lots shipped. After the root cause(s) of ES test failure are determined, corrected, and verified, the organization may resume shipments. The organization shall prevent shipment of suspect product without sorting or reworking, to eliminate the nonconformance. These ES requirements apply equally to sub-tier suppliers.

7.5.3–7.5.3.2 No changes required

7.5.3.2.1 Record retention

Inspection and Measurement Records: The organization shall retain records of process control data, product inspection data and records of appropriate reaction actions to readings outside the specification in a recoverable format for a minimum of 2 years, available to Ford Motor Company upon request. The organization shall record the actual values of process parameters and product test results (variable or attribute). Simple pass/fail records of inspection are not acceptable for variable measurements.

Audits: The organization shall retain records of internal quality system audits and management review for three years.

APQP: The organization shall maintain the final external supplier APQP/PPAP Readiness Assessment (Schedule A) for the life of the part (production and service) plus one year as part of the PPAP record.

Training: The organization shall retain records of training for 3 years from the date of the training.

Job setup: The organization shall retain records of job set-up verifications for 1 year. Retention periods longer than those specified above may be specified by an organization in its procedures.

Maintenance: The organization shall retain records of maintenance for 1 year. Also, the organization shall retain records of measurement equipment calibration for one calendar year or superseded, whichever is longer. Ford reserves the right to modify specific record retention requirements. These requirements do not supersede any regulatory requirements.

8 OPERATION

8.1 OPERATIONAL PLANNING AND CONTROL

Statement of Work: Appropriate to the organization's responsibilities, the organization shall meet the requirements of the statement of work. There may be an engineering statement of work (available from the Ford product development engineer), an assembly statement of work, a manufacturing statement of work or other types available from the appropriate Ford organization. See the global product development system (GPDS) for specific timing.

APQP: The external supplier APQP/PPAP READINESS ASSESSMENT (Schedule A) is available through https://web.qpr.ford.com/sta/APQP.html. The organization shall submit completed Schedule As as specified in the Schedule A notification letter for each program (monthly and after any significant change in APQP status). This applies to priority and non-priority suppliers, see Supplier Engagement Process on https://web.qpr.ford.com/sta/GPDSSupplierEngagement.html. Even if the organization has not received a Schedule A notification letter for a program but has new tooled end items (NTEIs) for a Ford program launch, the organization is still required to complete a Schedule A for each program milestone for all NTEIs and retain the final Schedule A in the PPAP file for the life of part (production and service) plus one year.

Prototypes: When the organization is also sourced with the production of prototypes, effective use should be made of data from prototype fabrication to plan the production process. The organization records the dimensional data per the PBCP, reviews the measured characteristics with Ford PD Engineer and obtains approval on the results from the Ford PD Engineer with confirmed acceptance of parts. If prototype parts are not fully compliant to specification, Ford PD engineering can approve use of the part with a WERS Alert. The organization should use the APQP/PPAP Evidence Workbook to record prototype part data for Ford PD review. The APQP/PPAP Evidence Workbook is available through https://web.qpr.ford.com/sta/APQP.html.

Prototype Tooling: Within 30 days of production verification (PPAP Phase 2) completion, the organization shall (i) complete the Prototype Disposal Request form, which can be obtained through a request to fordtool@ford.com; and (ii) submit the completed form to D&R supervisor for signature concurrence; (iii) submit signed form to fordtool@ford.com for processing.

8.1.1–8.2 No changes required

8.2.1 Customer communication

After part approval, the organization shall use the SREA (Supplier Request for Engineering Approval) process to submit approval requests for organization-initiated process change proposals. Refer to https://web.qpr.ford.com/sta/SREA.html.

8.2.1.1–8.2.3.1 No changes required

8.2.3.1.1 Review of the requirements for products and services—supplemental

The customer authorization for waiving formal review may be obtained from the appropriate Ford Organization (Ford engineering, purchasing, etc.).

8.2.3.1.2 Customer-designated special characteristics

Symbols: The organization is to contact Ford engineering to obtain concurrence for the use of Ford Motor Company special characteristics symbols defined as:

- Critical characteristic (CC or ∇ – inverted delta) (with safety or legal consideration, S = 9–10)
- Significant characteristic (SC) (Not Relating to Safety or Legal Considerations) S = 5–8)
- High impact (HI) characteristics
- Operator safety characteristics (OS)

For internal use, the organization may develop its own special characteristics symbols. The Special Characteristics definitions are available in the Ford FMEA Handbook (2011).

For fasteners, base part numbers beginning with "W9" are to be treated as inverted delta. *Critical characteristics* for fasteners may be designated by methods defined in the Ford Engineering Fastener Specifications available through the Ford Global Materials and Fastener Standards.

Other Special Characteristics: SC, HI, and OS characteristics are described in the Ford FMEA Handbook.

8.2.3.1.3 Organization manufacturing feasibility

Manufacturing feasibility reviews for updated or new manufacturing processes or capacity increases requiring tooling or equipment shall be documented as specified on the Manufacturing Feasibility form (both initial feasibility and final feasibility) https://web.qpr.ford.com/sta/Feasibility_Form.xls, per the timing specified on https://web.qpr.ford.com/sta/APQP.html and shall include all appropriate organization and Ford organizations.

8.2.3.2–8.3.1 No changes required

8.3.1.1 Design and development of products and services—supplemental

The organization should consider incoming inspection when developing control strategies to prevent the use of non-conforming incoming material.

8.3.2 No changes required

8.3.2.1 Design and development planning—supplemental

- *FMEA and CP Development*: FMEAs and CPs shall ensure that the manufacturing process complies with critical to quality process requirements as specified in the supplier manufacturing health charts located at https://web.qpr.ford.com/sta/Supplier_Manufacturing_Health_Charts.html. Approvals required for inverted delta parts.
- *Process FMEA(s) and CP(s)* for inverted delta component(s) require Ford engineering and STA approval in writing. Approvals required for all parts where the organization is design responsible.
- *Design FMEA(s)* prepared by design responsible organizations require Ford engineering approval in writing. Approval of revisions to these documents after initial acceptance per the above is also required. Ford reserves the right to require approval of FMEA and/or control plans for any part from any organization.
- *FMEA requirements*: Organizations shall comply with the Ford FMEA Handbook requirements; see FSP Document Library, https://fsp.portal.covisint.com/web/portal/document_library. Organizations complying with the Ford FMEA Handbook will meet the FMEA and related requirements of the Q1 3rd ed. MSA.
- *Families of FMEAs:* The organization may write FMEAs for families of parts, where typically the only difference in the parts is dimensional, not form, application, or function. The organization should obtain STA review and concurrence prior to use of family process FMEAs. The organization should obtain Ford PD review and concurrence prior to use of family design FMEAs.
- *FMEA documentation*: Organizations are to provide copies of FMEA documents to Ford Motor Company upon request. Special characteristic traceability for build to print organizations. For build to print organizations, the organization shall obtain from Ford DFMEA information (including potential CCs—YCs and potential Significant Characteristics (YSs) to develop the SCCAF, PFMEA and special characteristics (CC, SC, HI, and OS, as appropriate). The organization shall document special characteristics on the special characteristics communication and Agreement Form—SCCAF (FAF03-111-2) including where special characteristics are controlled at sub-tier suppliers and obtain Ford approval. The SCCAF template is

available through APQP/PPAP Evidence Workbook (through https://web.qpr.ford.com/sta/APQP.html). This also applies to Ford-directed sub-tier suppliers without a multi-party agreement.

- *Documentation of controls for critical characteristics*: Both build-to-print and design responsible organizations identify in the APQP/PPAP Evidence Workbook the special controls to prevent shipment of any nonconformance to Ford specified CCs, regardless of the location of the special controls in the supply chain (tier 1 through tier N).
- *Control plans*: All Ford Motor Company parts shall have CPs (or Dynamic Control Plans—DCPs, if required by Powertrain).
- *Special characteristic traceability*: Special characteristics and control approach are traceable from the DFMEA through the PFMEA and the SCCAF to the Control Plan and recorded in the APQP/PPAP Evidence Workbook.
- *Ongoing engineering specification testing documentation*: Any revisions to the product validation engineering specification or other inspection frequencies in the CPs and PFMEAs require Ford approval through the SREA.
- *Pre-launch control plans*: Pre-launch control plans shall be completed and utilized during production of parts from (TT)/(Unit TT) until final process capability approval is achieved.
- *Note*: The production control plan may be used for demonstration of Phase 3 with STA concurrence.
- *Submission of pre-launch control plan data*: Organizations providing parts to Ford Powertrain plants shall submit, to the Ford Powertrain Plant, the pre-launch control plan data for all (Unit TT) and (Unit PP) parts as specified by Ford.
- *Control Item* (∇) *Fasteners*: The following control shall be included in the CP for fasteners that are control items:
    - *Material analysis—heat-treated parts:* Prior to release of metal from an identified mill heat, a sample from at least one coil or bundle of wire, rod, strip, or sheet steel shall be analyzed and tested to determine its conformance to specifications for chemical composition and quenched hardness. The organization shall test a sample from each additional coil or bundle in the heat for either chemical composition or quenched hardness. The organization shall document the results and include the steel supplier's mill heat number. This requirement applies to both purchased material and material produced by the organization.

- *Material analysis—non-heat-treated parts*: The organization shall visually check the identification of each coil or bundle of wire, rod, strip, or sheet steel to determine that the mill heat number agrees with the steel supplier's mill analysis document and applicable specifications. The organization shall test each coil or bundle for hardness and other applicable physical properties.
- *Lot traceability*: The organization shall maintain lot traceability.

8.3.2.2–8.3.3.3 No changes required. For the special characteristics, see 8.2.3.1.2

8.3.4 Design and development controls

The organization shall perform design verification (DV) to show conformance to the appropriate Ford engineering requirements: attribute requirements list (ARL) and system design specification (SDS). The organization shall record the DV methods with the test results and submit to Ford product engineering for approval. For organizations responsible for component level DV testing, the organization shall have a documented Design Verification Plan and Report (DVP&R) that includes organization /sub-tier supplier and Ford responsible test(s) as applicable. The organization provides evidence of successful completion on all component level DV testing on the DVP&R. The organization shall obtain Ford PD engineer approval for all tests and results. These requirements apply to all organizations; regardless of the organization's or part's PPAP submission level or design responsibility. ARLs and SDSs are available from Ford product engineering. The organization shall use a GPDS when reviewing product design and development stages. Information on GPDS is available through the FSP (Ford Supplier Portal, https://fsp.covisint.com); log into FSP and then go to the Ford Supplier Learning Institute (FSLI) application.

Product development: For inverted delta (∇) parts, design responsible organizations shall include Ford engineering and assembly/manufacturing in GPDS milestone design reviews, as appropriate. Where feasible, design responsible organizations shall include Ford engineering and Ford assembly and/or manufacturing in design reviews for all Ford parts.

8.3.4.1–8.3.4.2 No changes required

8.3.4.3 Prototype program

The organization is responsible for the quality of the parts it produces and for any subcontracted services, including sub-tier suppliers specified by Ford Motor Company without a multi-party agreement. This applies to all phases of product development, including prototypes. Individual statements of work may specify alternate responsibilities. See the GPDS for additional information on prototype programs on the FSP.

8.3.4.4 Product approval process

PPAP: For production parts and approval of components from sub-tier suppliers, the organization shall comply with the AIAG: PPAP manual and Ford's Global Phased PPAP available through https://web.qpr.ford.com/sta/Phased_PPAP.html. Additional requirements are specified in Q1 3rd ed. at https://web.qpr.ford.com/sta/Q1.html. For service parts, in addition to meeting the requirements of the AIAG: PPAP manual, the organization must comply with the AIAG Service Production Part Approval Process (Service PPAP) manual.

Submission of sub-tier supplier PPAP: Evidence of sub-tier component part approvals may be a summary (approved PSWs, a listing of PSW approvals or equivalent).

Organization initiated changes: Per PPAP, the organization shall submit via WERS all organization-initiated design change proposals, unless the organization or sub-tier supplier does not have access to WERS. After SREA approval and change implementation, all changes require PPAP approval and functional trial approval or PPAP approval and functional trial waiver prior to shipping production quantities. STA will not grant full PPAP approval if the part or manufacturing process is under WERS Alert, per exception management process. See https://web.qpr.ford.com/sta/Phased_PPAP.html.

SREAs for service parts: The organization should process supplier-initiated change requests associated with service-unique parts no longer used in Ford production via the applicable Ford customer service department (FCSD) service part deviation SREA process found via https://web.srea.ford.com/ through the FSP. Contact your local FCSD STA engineer for further clarification.

8.3.5–8.4.1.1 No changes required

8.4.1.2 Supplier selection process

The organization's supplier selection process should include evaluation of the supplier's supply chain management system. The organization shall complete a financial assessment of the supply chain at a minimum annually, in conjunction with the annual audit program (see 9.2.2.2 of IATF 16949), not just at the initial supplier selection.

8.4.1.3 Customer-directed sources (also known as "directed-buy")

When required by the contract with Ford, the organization shall obtain approval from Ford Motor Company prior to sourcing sub-tier suppliers. Please, contact the Ford Buyer.

8.4.2 No changes required

8.4.2.1 Type and extent of control—supplemental

The organization shall have incoming product quality measures and shall use those measures as key indicators of sub-tier supplier product quality management.

8.4.2.2 Statutory and regulatory requirements

Applicable regulations shall include international requirements for export vehicles as specified by Ford Motor Company, e.g., plastic part marking (E-4 drafting standard—WSS-M99P9999-A1 and European End of Life of Vehicle (ELV)—available on the FSP. Material reporting requirements for ELV are specified by WSS-M99P9999-A1 under "Important Documents."

8.4.2.3 Supplier quality management system development

The organization may meet this requirement by successful assessments of the Sub-tier suppliers per the authorization stated on https://web.qpr.ford.com/sta/. The frequency of these reviews shall be appropriate to the sub-tier supplier impact on customer satisfaction. Sub-tier supplier quality management system requirements.

- Where a sub-tier supplier is not third-party certified to ISO/TS 16949 or IATF 16949, Ford reserves the right to require the organization to ensure sub-tier supplier compliance with the "Minimum Automotive Quality Management System Requirements for Sub-tier Suppliers" available through http://iatfglobaloversight.org/default.aspx. Evidence of effectiveness shall be based on having a defined process and implementation of the process, including measurement and monitoring.
- Where any organization has sub-tier suppliers not third-party certified to ISO/TS 16949 or IATF 16949, the organization is encouraged to require sub-tier supplier compliance

with the "Minimum Automotive Quality Management System Requirements for Sub-tier Suppliers."

- Ford or organization second-party assessment or third-party certification of sub-tier suppliers do not relieve the organization of full responsibility for the quality of the supplied product from the sub-tier supplier (including Ford-directed sub-tier suppliers without a multi-party agreement). Although all sub-tier suppliers must be assessed per this section, sub-tier supplier improvement efforts shall focus on those sub-tier suppliers with the highest impact on supplier improvement metrics (SIM).
- *Sub-tier supplier management process.* Organizations are encouraged to apply the principles outlined in "CQI-19 AIAG Sub-tier Supplier Management Process Guideline" to all their sub-tier suppliers.
- Additionally, Ford reserves the right to require the organization to apply the principles outlined in "CQI-19 AIAG Sub-tier Supplier Management Process Guideline" to address issues identified in the organization's supplier development and management process. Ford will communicate the requirement to apply CQI-19 to the specifically selected organization(s) based on sub-tier supplier management issues attributed to the organization. Evidence of effectiveness shall be based on having a defined process and implementation of the process including measurement and monitoring.

CC controls at the sub-tier suppliers: For CCs, the responsible organization ensures that sub-tier suppliers have controls in place to prevent shipment of non-conforming product at the location where the associated physical characteristics are manufactured by sub-tier suppliers. The sub-tier supplier controls for the CCs are identified by the organization in the APQP/PPAP Evidence Workbook. This also applies to Ford-directed sub-tier suppliers without a multi-party agreement.

8.4.2.3.1 No changes required

8.4.2.4 Supplier monitoring

In support of Ford's expectation of 100% on-time delivery, the organization shall also require 100% on-time delivery from sub-tier suppliers. The organization shall communicate any delay or risk to the affected Ford customer. The organization should monitor and minimize any premium freight expenses related to sub-tier suppliers for late deliveries. These also apply to Ford-directed sub-tier suppliers without a multi-party agreement.

8.4.2.4.1–8.5.1.1 No changes required

8.5.1.2 Standardized work—operator instructions and visual standards

Operators shall use the most current work instructions. The organization shall ensure that work instructions contain reaction plans for non-conformances showing the specific required steps.

8.5.1.3–8.5.1.7 No changes required

8.5.2 Identification and traceability

The organization shall meet all logistics requirements as specified by material planning and logistics (MP&L). MP&L requirements are available in the global terms and condition (GTC) web guides at https://web.fsp.ford.com/gtc/production/index.jsp?category=guides and on "MP&L-in-a-Box" at https://comm.extsp.ford.com/sites/MPLB2B/Pages/MPLdefault.aspx.

The organization is required to achieve level "A" on the Material Management Operation Guideline/Logistics Evaluation (MMOG/LE) to achieve and maintain Q1 3rd ed. Key requirements for MMOG/LE compliance include: (a) Annual MMOG/LE assessment completed and reported May 1 to July 31 each year; (b) adherence to Ford production and service delivery rating requirements for all regions as stated in Q1 3rd ed.; (c) part identification and tracking; (d) lot traceability throughout the value chain (lot traceability shall include subcontracted components of an assembly/module that are associated with compliance to any inverted delta requirement); (e) electronic communication with Ford and sub-tier suppliers; (f) management and maintenance of the Ford DDL CMMS3 system; (g) prevention of damage or deterioration of supplied products; (h) use of the appropriate packaging forms and maintenance of the Ford DDL CMMS3 DAIA Packaging screen, as applicable (packaging requirements and forms can be found in the packaging GTC Web Guides at https://web.fsp.ford.com/gtc/production/index.jsp?category=guides); (i) management and maintenance of returnable dunnage (Returnable container requirements are available through the GTC Web Guides at https://web.fsp.ford.com/gtc/production/index.jsp?category=guides); and (j) adequately trained personnel, as defined in MMOG/LE. In all cases, if unsure of the MP&L requirements, contact the production and service delivery analyst for the organization site, for each region. The analyst contact information is available through SIM.

Inverted delta part identification: The inverted delta symbol (∇) shall precede the Ford Motor Company part number for parts with CC, in accordance with the packaging guidelines for production parts and shipping parts and the identification label standard, both available through the FSP MP&L page, https://comm.extsp.ford.com/sites/MPLB2B/Pages/MPLdefault.aspx

> **Note:** Branding (E108) does not require the inverted delta symbol to be included with the part number physically marked on the part.

8.5.2.1–8.6 NO CHANGES REQUIRED

8.6.1 Release of products and services—supplemental

Ford reserves the right to require the use of an independent third-party inspector to ensure that the organization only ships compliant product to Ford facilities.

8.6.2 Layout inspection and functional testing

The organization shall perform annually a layout inspection (to all dimensional requirements) on at least 5 parts. Where tooling has multiple cavities, tools, or centers, the organization conducts the annual layout on at least one part from each cavity, tool, or center, with a minimum overall sample of 5 parts.

> **Note:** 5 parts are not required from each cavity, tool, or center, only a minimum of 1 part is required from each cavity, tool, or center. The measurements are to be documented on the APQP/PPAP Evidence Workbook (prototype or production measurement results section), available through https://web.qpr.ford.com/sta/APQP.html.

8.6.3 Appearance items

Appearance approval requirements are specified in PPAP, Ford CSRs, https://web.qpr.ford.com/sta/Phased_PPAP.html.

8.6.4–8.6.5 No changes required

8.6.6 Acceptance criteria

For guidance on product monitoring and reaction plan techniques for product conformance to specification, see the references AIAG: SPC and AIAG: APQP. For ongoing process capability requirements, see Table 7.2.

Table 7.2 Ongoing process capability requirements

The control chart indicates that the process:	Actions on the process output Based on process capability (Ppk)	
	Less than 1.33	Equal to or greater than 1.33
Is in control	100% inspect[a]	Accept product. Continue to reduce product variation
Has gone out of control	Equal to or greater than 1.33	Identify special cause 100% inspect[a] all product since the last in-control sample

[a] The organization ensures that the 100% inspection methodology prevents shipment of any non-conforming product to Ford. The 100% inspection methodology would typically include error proofing, such as a *poka-yoke*. The organization ensures that critical characteristics (CC) have controls that prevent the shipment of nonconforming product, regardless of the location in the supply chain (tier 1 through tier N) of the manufacture of the physical characteristic(s) associated with the CC. The organization records the CC controls in the APQP/PPAP Evidence Workbook. Statistical process control on product characteristics without continuous manufacturing process controls is not appropriate or sufficient for CC.

8.7–8.7.1 No changes required

8.7.1.1 Customer authorization for concession

Ford Motor Company authorization of product differing from Ford specifications is managed by WERS and is limited to the quantity of parts or time-period approved in the WERS Alert. This is applicable to both prototype and production level parts. PPAP submission and interim PSW acceptance are required for production use of parts with a WERS Alert. Where written by the organization, Alerts must contain the following:

- The specific PPAP requirements that are not completed.
- The modified specifications(s) that the part satisfies.
- The justification why the modified specification(s) is acceptable.
- The containment plan to assure the quality of parts (e.g., extraordinary controls, inspection process, robust measurement systems).
- The period (typically in terms of days), the number of parts and the specific launch build event for which the Alert is effective. The WERS help desk can provide information on WERS via email: hwers@ford.com. WERS training is available through FSLI through https://fsp.portal.covisint.com/web/portal/home.

8.7.1.2–9 NO CHANGES REQUIRED

9.1 MONITORING, MEASUREMENT, ANALYSIS AND EVALUATION

Ford reserves the right to request the data collected by the organization as defined in either the pre-launch or production CPs.

9.1.1 No changes required

9.1.1.1 Monitoring and measurement of manufacturing processes

Table 7.2 shows the ongoing process capability requirements. All process controls shall have a goal of reduction of variability, using P_{pk}, six sigma, or other appropriate methods. Any statistical process control book or manual can provide additional guidance where tool wear impacts variability (See AIAG: SPC, 2005; Stamatis, 2002 v.4; Wheeler, 2010 and many others). All process metrics are to be traceable to Ford requirements.

9.1.1.2 Identification of statistical tools

The organization shall use the latest edition of the following references as appropriate: See IATF 16949 for applicable references.

Process Capability: The capability index for reporting launch process capability and ongoing production process capability is P_{pk} (Performance Index). See Ford's PPAP customer specifics for the launch process capability requirements, https://web.qpr.ford.com/sta/Ford_Specifics_for_PPAP.pdf. The organization shall maintain ongoing process capability at $P_{pk} > 1.33$. The requirement for maintenance of ongoing process capability is to be included in the production CP and the capability results recorded in the APQP/PPAP Evidence Workbook. The results of monitoring process capability are to be available to Ford upon request. When investigating a process capability issue it is advisable to use multiple indices, e.g., P_p, P_{pk}, C_p, and C_{pk}. When used together, the indices assist in the determination of sources of variation (See AIAG: SPC, 2005; Stamatis, 2002 v.4; Wheeler, 2010).

9.1.1.3 No changes required

9.1.2 Customer satisfaction

The organization shall monitor performance and customer satisfaction metrics (as defined by Q1 3rd ed.) and updates to Ford requirements on the FSP. The control chart indicates that the process is in control, consistent and predictable:

Identify special cause: It is recommended that the organization review their performance status on SIM at least weekly (some information is updated daily in SIM). At least twice per year, the organization shall communicate customer satisfaction metrics to all employees who affect the quality of Ford Motor Company parts.

9.1.2.1 Customer satisfaction—supplemental

Certification body notification: The organization shall notify its certification body of record in writing within five (5) working days if Ford Motor Company places the site on Q1 3rd ed. revocation. This notification of the certification body will constitute a "customer claim" as defined by the ISO/TS 16949/IATF 16949 Rules. This step will suspend the organization's ISO/TS 16949/IATF 16949 certification. However, a suspended certification is still acceptable for Q1 3rd ed. capable systems requirements. Even though the certification body may request a status report from Ford on the implementation of corrective actions to address the Q1 3rd ed. revocation, the certification body alone must determine whether to withdraw the ISO/TS 16949 certification within 90 days of suspension.

Note 1: Reinstatement of Q1 3rd ed. from revocation requires at least 6 months of acceptable organization performance—in some instances 12 months. (In reality however, most suppliers will get their recertification in approximately 18–24 months.) The certification body may remove suspension of the ISO/ TS 16949/ IATF 16949 certificate if the organization's corrective actions have addressed the non-conformances leading to the certificate suspension, as determined by the certification body. The certification body may remove the suspension even though the site remains under Q1 3rd ed. revoked status, accumulating the required 6 months of acceptable performance data.

Note 2: At its option, Ford may file an OEM performance complaint with a certification body when confronted with a specific organization quality performance issue where a root cause may be a nonconformance in the organization's quality management system. Ford will send the notification letter to both the organization and the certification body's oversight office.

9.1.3–9.2.2.2 No changes required

9.2.2.3 Manufacturing process audit

Ford Manufacturing Process Assessment Requirements: The organization is responsible to ensure that all tiers of suppliers are assessed to the applicable Ford manufacturing process standards.

Note: Self-assessment by the sub-tier suppliers, including implementation of corrective action plans as required, meets this requirement. Refer to https://web.qpr.ford.com/sta/Ford_GTS.html on the FSP for all these standards except AIAG CQI-xx, which are available through AIAG.

Ford supplier manufacturing health chart requirements: The organization shall assess compliance to critical to quality process requirements in accordance with APQP as specified in the supplier manufacturing health charts located at https://web.qpr.ford.com/sta/Supplier_Manufacturing_Health_Charts.html.

Heat treat assessment requirements: Organizations and sub-tier suppliers providing heat-treated product and heat-treating services shall demonstrate compliance to AIAG CQI-9 "Special Process: Heat Treat System Assessment" and Ford Specific CQI-9 requirements (available through https://web.qpr.ford.com/sta/CQI-9_Ford_Specific_requirements.xls); CQI-9 is available through AIAG, http://www.aiag.org/

- *CQI-9 Special Process*: heat treat system assessment. All heat-treating processes at each organization and sub-tier supplier manufacturing site shall be assessed annually (at all tier levels), using the AIAG CQI-9 "Special Process: Heat Treat System Assessment" (HTSA) and Ford.
- *Specific CQI-9 requirements*: Assessments are also required following any heat treat process and/or changes of heat-treat equipment or additions. The organization must review that the individual assessments are current (less than 12 months old), meet the requirements above and enter the CQI-9 assessment status into GSDB Online. The organization shall maintain the 2-prior annual CQI-9 assessment reports and related information at the organization's site and make them available to STA upon request. Heat treat assessments are conducted by the organization, heat treat suppliers, sub-tier suppliers, or

by Ford. Demonstration of compliance to CQI-9 and Ford Specific CQI-9 requirements does not relieve the organization of full responsibility for the quality of supplied product. To reduce the risk of embrittlement, heat-treated steel components shall conform to the requirements of Ford Engineering Material Specification WSS-M99A3-A.

9.2.2.4–9.3.1 No changes required

9.3.1.1 Management review—supplemental

The organization management shall hold monthly QOS performance meetings as specified in the Q1 MSA available on https://web.qpr.ford.com/sta/Q1.html. The results of these QOS performance reviews shall be integral to the senior management reviews.

Note: the organization need not hold the management review as one meeting, but it may be a series of meetings, covering each of the metrics monthly.

9.3.2 Management review inputs

Management review input must also include the Q1 MSA results.

9.3.2.1–10.2 NO CHANGES REQUIRED

10.2.1 and 10.2.2

The organization shall have processes and systems in place to prevent shipment of non-conforming product to any Ford Motor Company facility. The organization should analyze any non-conforming product or process output using the 8D methodology to ensure root cause correction and problem prevention.

Customer Concerns

Organizations shall respond to quality rejects (QRs) by:

- Responding in 24 hours.
- Implementing containment in the Ford plant. The organization and/or third party must follow local procedures and site rules while carrying out containment.
- Providing certified stock.
- Delivering an 8D, beginning with symptom and emergency response actions (D0) through interim containment actions (D3).

- Within 48 hours of notification by the Ford plant, the organization shall notify Ford Service if the product quality issue is suspected of affecting any FCSD shipments.
- Within 15 calendar days, delivering the 8D or (six sigma) six panel with preliminary or verified root cause, and a plan to implement corrective and preventive actions with supporting data.

Note: The clock starts once Ford has sent the notification to the organization.

A summary of the quality reject process for North America is available through https://web.qpr.ford.com/sta/QR2NA.htm. Global 8D system is available on FSP (https://web.quality.ford.com/g8d/).

Returned product test/analysis: The organization shall have a documented system for internal notification, analysis, and communication of all Ford plant returns and warranty returned parts.

- The organization shall communicate the results of analysis to the responsible Ford and organization work groups and include the results in the associated 8D report.
- The organization shall communicate Ford plant PPM (parts per million) to all organization plant team members.
- The organization shall develop a system to monitor Ford plant and warranty concerns. The organization shall also implement corrective actions to prevent future Ford plant and warranty concerns. Returned product test results are to be included in the monthly QOS performance report as part of the management review.

10.2.3 No changes required

10.2.4 Error-proofing

Commodity Specific (Internal Ford Stamping Facilities)

Internal stamping suppliers shall meet the requirements of Ford Motor Company, Vehicle Operations' Operating Procedure VOPQUN-050, "Production Part Approval Process (PPAP)/Part Submission Warrant (PSW) Process." A summary of submission practices and methods for notifying the appropriate Ford system of completion of the PSW is shown in Table 7.3:

Table 7.3 PPAP submission practices

PPAP Submission practice		
DDL (Direct Data Link)		
Level 1	**Level 3 or 5**	**Level 2 and 4 not used**
Non DDL • Prepare PPAP data package/ self-approve. • Enter approval in Ford system. • Maintain the completed PPAP data package on file. • Prepare PPAP data package/ self-approve. • Contact Ford STA and report PPAP data package approval • Maintain the completed PPAP data package on file. • Ford signature not required. • Organization self certifies. • Ford signature not required. • Organization self certifies. • Organization enters in Ford system. • Purchasing support function records approval, per local practice.	Non-DDL • Prepare PPAP data package. • Get STA approval on paper PSW (for priority supplier PD approval also required). • Enter approval in Ford system after STA (and PD for priority parts) approval of PSW and PPAP data package. • Maintain the completed PPAP data package on file. • Prepare PPAP data package. • Get STA approval on paper PSW (for priority supplier PD approval also required). • Maintain the completed PPAP data package on file. • Contact the purchasing support function and report PPAP data package approval status. • Approve PPAP data package/sign warrant • Return PPAP data package to organization. • Approve PPAP data package/sign warrant • Return PPAP data package to organization • Organization enters approval in Ford system after STA (and PD for priority parts) approval of PPAP data package. • Local practice (purchasing support function) records approval per local practice.	

Note: For DDL enabled organizations, PPAP status is submitted to Ford via the VPP (Vehicle) and MPP (Powertrain) systems.

Launching considerations—see Table 7.4

Table 7.4 Launching considerations

Launching considerations		
Key areas simplified	Key requirements deleted	Key enhancements
The following areas of the Ford Customer Specifics to IATF 16949 has been simplified compared with the corresponding requirements for ISO/TS 16949: • Quality operating system metrics • Measurement systems analysis • Location of Ford engineering standards • Record retention requirements • Control plan requirements • Sub-tier supplier assessment • Process capability • Heat treat submission and ratings	Examples of requirements that have been deleted because they are now incorporated into IATF 16949: • Lean manufacturing • Preparation of FMEAs • Focus on prevention • Sub-tier supplier part approval • Predictive and preventive maintenance • Internal auditor training	Two areas have been enhanced, clarifying new requirements in the IATF 16949 standard: • Risk management • Corporate responsibility

Sources:

Ford Motor Co. (May 1, 2017). *Ford Motor Company Customer-Specific Requirements for IATF-16949:2016.*

Ford Motor Co. (June 2013). *Ford Motor Company Customer-Specific Requirements for use with PPAP 4.0*

Ford Motor Co. (March 22, 2017). *Launch of Ford Customer Specifics to IATF 16949.*

Ford specific requirements material is http://www.iatfglobaloversight.org/wp/wp-content/uploads/2016/12/Ford-IATF-CSR-for-IATF-16949-1May2017.pdf. Retrieved on May 14, 2018.

Note: All Internet sites and procedures referenced in this document may be also found in the Covisint system. In order to get into the system, one must have their own ID and be authorized by Ford, but all the standards mentioned here may be found in the open Internet web.

FCA US LLC Customer-Specific Requirements for IATF 16949

Generally, the FCA specific requirements fall into these categories:

- Part-specific requirements (dimensions, materials, performance characteristics, etc.)
- Special characteristics
- Delivery requirements
- Boilerplate requirements (typically found in the purchase order)
- General requirements (PPAP, APQP, etc.)
- Process requirements (example: heat treat)

A more detailed description follows below.

4–7.2.1 NO CHANGES REQUIRED

7.2.2 Competence—on-the-job training

Each location shall have a sufficient number of trained individuals such that computer applications necessary for direct support of FCA US manufacturing can be accessed during scheduled FCA US operating times, and other applications can be regularly accessed during normal business hours. The specific computer applications required will vary with the scope of an organization site's operations. For manufacturing sites, the recommended quality applications include, but are not limited to: (a) 3CPR—3rd Party Containment and Problem Resolution; (b) CQMS—Corporate Quality Management System; (c) CQR—Common Quality Reporting; (d) DRIVe—Delivery Rating Improvement Verification; (e) EBSC—External Balanced Scorecard; (f) EWT—Early Warranty Tracking; (g) GIM—Global Issue Management; (h) GCS—Global Claims System; (i) NCT—Non-Conformance Tracking; (j) PRAS—Parts Return Analysis System; (k) QNA—Quality Narrative Analyzer; (l) webCN—Change Notice System; and (m) WIS—Warranty Information System.

Note: FCA US periodically offers training to organization personnel on selected FCA US processes and procedures (including those referenced in this document) during supplier training week. Information on content, scheduling and registration is available in the "Supplier Learning Center" application in eSupplierConnect.

7.2.3–7.3.2 No changes required

7.4 COMMUNICATION

Forever requirements: The organization shall comply with the forever requirements activities described in Appendix E of the PPAP Manual.

7.5–7.5.3 NO CHANGES REQUIRED

7.5.3.2.1 Record retention

Organization-controlled records: Records identified by FCA US as "organization-controlled" shall be retained on-site but made available for review by FCA US or the certification body upon request.

Minimum retention requirements: Quality performance records (e.g., control charts, inspection and test results) shall be retained for one calendar year after the year in which they were created. Records of internal quality system audits and management review shall be retained for three years.

7.5.3.2.2–8.2.1 No changes required

8.2.1.1 Customer communication—supplemental

The organization shall establish a connection for electronic communication with FCA US through eSupplierConnect.

Note: Instructions for registering for the portal and assistance with its use can be found at https://fcagroup.esupplierconnect.com.

8.2.2–8.2.3.1.1 No changes required

8.2.3.1.2 Customer-designated special characteristics—The Shield <S>; also <E>

The *Shield* identifies special characteristics that require special due diligence since the consequence of a likely assembly or manufacturing variation may cause a non-conformance to safety and regulatory product requirements. Suppliers (if applicable) shall be knowledgeable of the following standards: PFSAFETY<S>, PF-Emissions <E>. <S> designates product safety and/or regulatory requirements. <E> designates government-regulated vehicle emissions requirements.

The *Diamond* <D>. The Diamond identifies special characteristics of a component, material, assembly or vehicle assembly operation that are designated by FCA US as key to the function and customer acceptance of the finished product. Diamonds also highlight important characteristics on fixtures and gauging procedures during design verification, product validation, or revalidation. The symbol <D> identifies key but non-safety and non-regulatory product characteristics or processes that may be susceptible to manufacturing variation and require additional controls to assure conformance to specifications and customer satisfaction. A Diamond <D> requires that a process control plan be developed for that characteristic

Note: The use of a Diamond as specified in PS-7300 does not automatically require the use of statistical process control. Other methods of control (such as error-proofing and mistake-proofing) may be more able to prevent or detect nonconformances. Processes that demonstrate a high degree of capability (C_{pk} > 3.0, e.g.) for an extended period of time may require a less frequent method of control. The exact method to be used must be determined in advance and agreed to by the FCA US supplier quality engineer and product engineer. Presence of a Diamond does not affect the significance to a Shield(s) on the same document. For further detail, organizations shall refer to PS-7300.

Special characteristics not identified with symbols: FCA US or the organization may choose product or process characteristics that affect fit, form, function or appearance that are not identified with a symbol. Situations where this may occur and the applicable FCA US engineering standards addressing these situations are summarized in Table 7.5:

The organization may develop its own special characteristics symbols for internal use. If organization specific special characteristics are developed, the organization shall document the equivalence of the internal symbols with FCA US symbols and reference the equivalence when the organization uses internal symbols in its communications with FCA US.

Table 7.5 FCA US engineering standards addressing special characteristics not identified with symbols

If the organization	The organization should refer to
Provides engineering (including service) or assembly services, parts or components for vehicles intended for sale in regulated markets outside of NAFTA	PF-Homologation
Provides parts or components: (a) that require tracking to ensure emission, certification and regulatory compliances and (b) that are designated as high-theft components for law enforcement need	PS-10125
Provides appearance items—parts or components whose color, gloss, or surface finish requirements are specified by the FCA US Product Design Office	AS-10119

8.2.3.1.3–8.3.2 No changes required

8.3.2.1 Design and development planning—supplemental

FCA US uses the process planning review (PPR) and process audit (PA), documented in the process planning and audit tool, for documentation of advance quality planning. When required, organizations shall participate in teams to develop parts or components and shall use PPR and PA. On occasions when use of PPR and PA is not required, organizations shall develop products according to the advanced product quality planning (APQP) Process.

> **Note:** FCA US and FCA Italy SpA share common advance quality planning methods. An FCA US-led process planning review/process audit (PPR/PA) program shall be performed for parts that have a customer-monitored (high or medium) initial risk as identified by the supplier quality engineer. Supplier-monitored (low risk) parts shall have an organization-led program, unless otherwise specified by the FCA supplier quality engineer. Parts that have been out of production for 12 months or more shall have an organization-led PPR/PA unless otherwise determined by the supplier quality engineer. PPR/PA shall be completed prior to providing PS-level parts to FCA US and shall be completely approved prior to a PPAP submission. Unless otherwise specified, changes made to advance quality-planning processes are not retroactively applied to existing product development programs. In the absence of specific direction by FCA US, the organization shall implement quality management system changes in time to be in conformance during their next new product development program.

8.3.2.2–8.3.4.1 No changes required

8.3.4.2 Design and development validation

DV and PV shall be satisfactorily completed before PA and PPAP approval.

Note: Guidance on the extent of required PV testing is provided by the PPR/PA tool *Production Validation Testing Scope.*

8.3.4.3 No changes required

8.3.4.4 Product approval process

Process audit—A systematic and sequential review of the organization's process shall be completed through a process audit (PA) performed by the FCA supplier quality engineer and product engineer prior to a PPAP submittal. The purpose is to verify the organization's process readiness and to assure understanding of complete program requirements.

Production Part Approval Process—The organization shall comply with AIAG's *Production Part Approval Process (PPAP), 4th Edition, Service; Production Part Approval Process (Service PPAP), 1st Edition* and *FCA US Customer-Specific Requirements for Use with PPAP 4th Edition.*

8.3.5–8.3.5.1 No changes required

8.3.5.2 Manufacturing process design output

PFMEAs and control plans are required for prototype, pre-launch, and production phases. PFMEA and CP documentation shall be audited to the PFMEA and CP document audit form. CP shall be verified to the CP process audit checklist, with corrective action for any identified nonconformance(s) documented on the associated PDCA planning worksheet. An FCA US representative's signature is not required on CPs, unless specifically requested by the supplier quality engineer.

8.3.6–8.4.1.1 No changes required

8.4.1.2 Supplier selection process

With respect to suppliers to the organization ("sub-tier suppliers"), the organization shall:

- Conduct an on-site process audit (or equivalent) and production demonstration run (PDR) for all parts/suppliers that are NOT considered by FCA US or the organization to be low risk to the vehicle program.

- Develop and maintain a list of approved suppliers for each sub-component, raw material, commodity, technology, or purchased service that is not consigned or directed by FCA US. The organization shall have a documented process and use assigned personnel to monitor and manage performance.

8.4.1.3 Customer-directed sources (also known as "Directed–Buy")

If the organization has one or more directed parts/suppliers:

- FCA US is responsible for the process planning review, process audit, and PDR activities up to and including PPAP, with input from and participation of the organization (Tier 1 Supplier).
- The organization (Tier 1 Supplier) is responsible for managing the ongoing quality of the Tier 2 components following PPAP and working with FCA US to resolve issues.

If the organization has one or more consigned parts/suppliers, FCA US is responsible for all quality activities up to and including PPAP, as well as management of ongoing quality issues.

8.4.2–8.4.2.2 No changes required

8.4.2.3 Supplier quality management system development

Management and development of supplier QMS effectiveness shall be evaluated on the basis of evidence that the organization has processes in place that include such elements as:

  - Supplier QMS development strategy (8.4.2.5)
  - Criteria for designating "exempt" suppliers
  - Criteria for granting waivers to select suppliers for compliance to specified elements of ISO 9001 or IATF 16949
- Second-party audit administration (8.4.2.4.1)
  - Identification of second-party auditors
  - Criteria for granting self-certification status to qualified suppliers
  - A schedule for second-party audits
- Organization-controlled record keeping (7.5.3.2.1)
- Progress monitoring

Note: Organizations requiring additional guidance on supplier QMS development should refer to *CQI-19: Sub-tier Supplier Management Process Guideline* or minimum automotive quality management system requirements for sub-tier suppliers (MAQMSR). The organization shall prioritize the QMS development program for non-exempt suppliers to introduce compliance to the MAQMSR as the first step beyond compliance with ISO 9001 or certification to ISO 9001. At a minimum, the organization should require their non-exempt suppliers to demonstrate compliance to ISO 9001 and MAQMSR.

Ship-Direct Suppliers. Organizations may, with FCA US purchasing concurrence, identify a supplier location within FCA purchasing systems as an organization manufacturing site. (Such a designation allows direct shipment of manufactured goods to FCA US.) Unless otherwise specified by FCA US, such sites shall be subject to the registration requirements described in Section 1.2. In the event that FCA US chooses to grant such a supplier site an exemption to IATF 16949 registration,

- The site shall receive the highest priority for QMS development.
- The site shall not be designated "exempt" nor shall a waiver be granted without the written concurrence of FCA US Supplier Quality (supplier development not required of suppliers certified to IATF 16949). Supplier QMS certification by an IATF-recognized certification body to IATF 16949 completely satisfies the requirements for QMS development. Further QMS development by the organization is not required while the supplier's certification is valid. If the supplier certification expires or is cancelled or withdrawn by their certification body, the organization shall establish and implement a plan for second-party audits to ensure continued compliance to IATF 16949 until such time as the supplier is recertified. Exemption shall not be granted as an alternative to recertification without approval from FCA US Supplier Quality management.

8.4.2.3.1–8.4.2.4 No changes required

8.4.2.4.1 Second-party audits

Second-party audit administration. The second-party must annually audit each non-exempt supplier for whom it has performed the second-party service.

1. For suppliers not certified to ISO 9001, the duration of these audits must conform to the full application of the audit day requirements of the *Rules*, Section 5.2.
2. For ISO 9001 certified suppliers, audit length may vary to suit individual supplier requirements and audit resource availability in accordance with the documented development strategy. Audit reports shall be retained as organization-controlled records (7.5.3.2.1). The following second-party qualifications shall apply:
   - The organization must be certified to IATF 16949:2016 by an IATF-recognized certification body.
   - The IATF 16949 certification of the second party cannot be in "suspended" status.

Supplier self-certification: The organization shall have a documented process for identifying and qualifying suppliers for whom self-certification is an effective alternative to second-party audits for QMS development. Qualification criteria shall include a preliminary evaluation (audit) of the supplier's QMS, an analysis of the supplier's quality performance and an assessment of the incremental risk to organization products. Self-certification qualifications shall be documented and subject to periodic review. Such documents shall be managed as organization-controlled records (7.5.3.2.1).

8.4.2.5 Supplier development

Supplier exemptions/waivers. The organization strategy for supplier development of its active suppliers shall include a documented process for designating "exempt" suppliers—those suppliers who are unable or unwilling to fully certify a quality management system to IATF 16949 or ISO 9001. The organization development strategy shall include provisions for granting partial exemptions ("waivers") to suppliers providing commodities for which specific sections of ISO 9001 or IATF 16949 do not apply.

Except as noted in Section 8.4.2.3, declaring a supplier "exempt" does not relieve the organization of the responsibility for supplier QMS development for any sections of ISO 9001 or IATF 16949 not explicitly waived. Supplier development prioritization, exemption, and waiver decisions, as well as the scope of individual exemptions or waivers, shall be documented and subject to periodic review. This documentation shall be retained as an organization-controlled record.

8.4.3 No changes required

8.4.3.1 Information for external providers—supplemental

With respect to external providers to the organization (i.e., "sub-tier suppliers"), the organization shall:

- Cascade and communicate all FCA US quality requirements (e.g., quality planning, process audit, PDR, forever requirements, etc.) throughout the organization's supply chain.
- Initiate a forever requirement notice for any proposed process change throughout the supply chain.

8.5–8.5.4 No changes required

8.5.4.1 Preservation—supplemental

Organizations shall be familiar and comply with FCA US packaging, shipping and labeling requirements contained in the *Packaging and Shipping Instructions* manual.

8.5.5–8.6.1 No changes required

8.6.2 Layout inspection and functional testing

Annual Layout: To ensure continuing conformance to all FCA US requirements, a complete annual layout inspection, including all sub-components, shall be required for all production parts and components unless waived in writing by the FCA US supplier quality engineer. Any such waiver shall be subject to annual review and renewal. Documented evidence of the waiver shall be retained as an organization-controlled record. The frequency and extent of layout inspections for service parts and components shall be established by the organization with the written approval of Mopar® Supplier Quality. Documented evidence of the approved layout inspection plan shall be retained as an organization-controlled record. In the absence of a written agreement, an annual, full layout inspection is required.

8.6.3 Appearance items

Appearance master samples: All appearance masters are specified and controlled by the FCA US Product Design Office. Samples of appearance masters are available from the Thierry Corporation: http://www.thierry-corp.com.

8.6.4–9.1.1.3 No changes required

9.1.2 Customer satisfaction

External balanced scorecard: FCA US purchasing and supplier quality use the external balanced scorecard (EBSC) to evaluate customer satisfaction with its external production and service suppliers. EBSC stores, analyzes, and reports organization performance data collected from other sources within FCA US. The EBSC report used for evaluation of organization site performance at a commodity level is the monthly supplier scorecard ("scorecard"). The production scorecard reports ratings in five categories:

1. Incoming material quality (IMQ)
2. Delivery
3. Warranty
4. Cost
5. Partnership

The service and prototype scorecard reports performance in three categories:

1. Incoming material quality (IMQ)
2. Delivery
3. Partnership

Cost and partnership are used to measure commercial performance and shall not to be used to evaluate the performance of organization quality management systems. Supplier quality reporting FCA US may, at its discretion, provide certification bodies with periodic reports of their clients' quality data, such as:

- EBSC IMQ, delivery, and warranty metrics with supporting data.
- FCA US supplier quality process audit reports.

Note: Sharing certification body client quality data does not constitute an OEM performance complaint as described in Section 8.1 of the *Rules*.

9.1.2.1 *Customer satisfaction—supplemental*

OEM Performance Complaint. FCA US may, at its discretion, file an OEM performance complaint with a certification body when confronted with a specific organization quality performance issue where a root cause may be a nonconformance in the organization's quality management system. FCA US shall notify the certification body of the OEM performance complaint by sending the certification body a notification letter that will: (a) identify the organization site, (b) summarize substance of the complaint, (c) document the affected element(s) of IATF 16949, and (d) request a copy of the organization site's last audit report.

> **Note:** As FCA US is an IATF member; a request for client audit reports is permitted under Section 3.1.e of the *Rules.* A copy of the notification letter will be sent to the organization, as well as the certification body's oversight office. Upon receipt of the OEM performance complaint notification letter, the certification body shall investigate the complaint in accordance with Section 8.0 of the *Rules.* At the conclusion of their investigation, the certification body shall advise FCA US of their findings and any actions taken. An OEM performance complaint may be filed in conjunction with, or independently of, a top problem supplier location (TPSL) action. The certification body findings from an OEM complaint investigation may be used by FCA US to establish the need to place an organization site in TPSL or new business hold.

Top problem supplier location reporting. Upon periodic review of EBSC quality measures and other key performance indicators, FCA US may notify specific organization sites that they have been identified as a TPSL. The TPSL designation signals FCA US dissatisfaction with the organization site's quality performance and begins a process to develop and implement a performance improvement plan.

FCA US shall notify the certification body of the organization site's involvement in the TPSL process by sending the certification body a copy of the notification letter and follow-up communications (as required) that will: (a) identify the organization site, (b) summarize the process, (c) document specific areas of concern with supporting data, and (d) request a copy of the organization site's last audit

Note: As FCA US is an IATF member; a request for client audit reports is permitted under Section 3.1.e of the *Rules*. Certification body notification of TPSL activity is for information only and does not constitute an OEM performance complaint as described in Section 8.1 of the *Rules*. However, FCA US reserves the right to file a performance complaint at any point within the TPSL process.

FCA US shall notify the certification body when the organization site has achieved the agreed-upon exit criteria and is removed from the TSPL process. Quality new business hold upon periodic review of EBSC quality measures and other key performance indicators, FCA US may notify an organization that they have been placed in quality new business hold (QNBH) status. This indicates that the organization site's quality performance is persistently below expectations and corrective action is required. The organization will be ineligible to bid on new FCA US business supplied from the affected organization site(s) without senior purchasing management intervention. A notification letter is sent to the organization, outlining the substance of the complaint and identifying the exit criteria the organization must achieve to be removed from QNBH status. A separate notification letter is sent to the organization's certification body and the oversight office via electronic mail. This letter will: (a) identify the organization, (b) describe the substance of the complaint, (c) provide evidence supporting the complaint (the organization notification letter and additional data as required), and (d) identify the FCA US Supplier Quality representative for the complaint. The certification body shall: (i) issue a major nonconformance against the organization and suspend the organization's ISO/TS 16949 certification in accordance with Section 8.0 of the *Rules*, and (ii) provide FCA US with copies of the organization's last recertification audit and all subsequent surveillance audits.

Note: As FCA US is an IATF member; a request for client audit reports is permitted under Section 3.1.e of the *Rules* (a) Follow the process outlined in Section 8.0 of the *Rules* to manage the nonconformance and determine whether the organization's certificate will be restored or withdrawn. If the certification body reinstates the organization's certificate, the organization will remain in QNBH status beyond the reinstatement date while FCA US monitors EBSC quality measures and other key performance indicators. If the effectiveness of the implemented corrective actions cannot be verified, FCA US shall refer the issue to the organization's certification body and their oversight office for further investigation. The organization site shall remain in NBH status. When the exit criteria established for the organization have been met, FCA US shall: (i) remove the QNBH status, lifting the associated commercial and quality sanctions (Sanctions imposed by other FCA US processes may remain in place) and (ii) notify the affected organization site(s), the certification body and the oversight office. If the certification body withdraws the certificate, FCA US purchasing and supplier quality management will develop a joint plan for the organization that either restricts further commercial activity or works toward improving processes and performance to a level that permits the organization to petition for new certification. If an organization site is seeking certification to ISO/TS 16949 or IATF 16949, but is placed on QNBH status before the stage 2 audit is conducted, the certification body shall not conduct a stage 2 audit until the QNBH status is lifted or FCA US supplier quality management notifies the organization and the certification body in writing that the stage 2 audit may proceed. If an organization site is placed on QNBH status after a stage 2, transfer or recertification audit, but before the certificate is issued: (i) the certification body shall immediately suspend the existing certificate (if applicable), (ii) the certification body shall issue the new certificate in accordance with the *Rules*, and (iii) the certification body shall then immediately place the new certificate in suspension in accordance with the *Rules*. If applicable, the suspension of the previous certificate shall be removed.

Material Management Operations Guideline/Logistics Evaluation (MMOG/LE): Organizations shall use Global MMOG/LE–Version 4 to integrate evaluation of delivery performance into their quality management system. Evaluation of integration effectiveness shall be based on evidence that the organization has a process in place that includes elements such as: (a) internal auditors identified; (b) an established schedule for self-assessment (including evidence of schedule adherence); (c) timely submission of the completed self-assessment to FCA US; (d) a defined continuous improvement process (including evidence of goal-setting and performance evaluation); (e) a defined corrective action process (including evidence of actions taken and verification of effectiveness); and (f) progress monitoring. Evaluation shall be by self-assessment. The self-assessment shall be conducted annually but may be repeated as needed.

> **Note:** FCA may choose to conduct a MMOG/LE audit at any time. The self-assessment shall be conducted using the "Full" self-assessment spreadsheet tool from Global MMOG/LE–Version 4. The results of the annual self-assessment shall be submitted to FCA US through the DRIVe system (accessible through eSupplier-Connect) between May 1 and July 31 of the current calendar year. A copy of the completed spreadsheet shall be retained. Questions concerning MMOG/LE should be directed to FCA US supplier delivery development at scmsdd@fcagroup.com.

9.1.3 No changes required

9.2.2.2 Quality management system audit

The scope of the annual audit program shall include a review of a minimum of two product control plans for FCA US parts, where applicable.

9.2.2.3 Manufacturing process audit

Layered Process Audits: Organizations supplying production parts or components to FCA US shall conduct layered process audits (LPA) on all elements of manufacturing and assembly lines that produce production parts or components for FCA US. These shall include both process control audits (PCA) and error-proofing verification (EPV) audits. Organizations shall provide evidence of compliance to the following requirements:

- Audit process shall involve multiple levels of site management, from line supervisor up to the highest level of senior management normally present at the organization site. A member of site senior management shall conduct process control audits at least once per week. All members of site senior management shall conduct process control audits on a regular basis.
- Delegation of this activity will not be accepted, with the exception of extenuating circumstances.
- The organization shall have a documented audit structure with auditor level and frequency of inspection.
- PCAs shall be conducted at least once per shift for build techniques and craftsmanship related processes.
- EPV audits shall be conducted at least once per shift, preferably at the start of shift. Compliance charts shall be completed once

per quarter and maintained for the life of the program. The following metrics shall be included:
  - Audit completion by all auditing layers
  - By-item percentage conformance by area

- Reaction plans shall be in place to immediately resolve all non-conformances. The organization shall show evidence of immediate corrective action, containment (as required), and root cause analysis (as required). A separate communication procedure is required to address reoccurring non-conformances. Specific areas of focus shall include the following:
  - Resolution of non-conformances
  - Escalation of issue for management review
  - Lessons learned

LPAs are not required for specific materials, parts, or assemblies produced on such an infrequent or irregular basis that it would prohibit establishing a regular, weekly audit schedule.

- Such infrequently or irregularly produced materials, parts, or assemblies shall be subject, at a minimum, to a process audit at start-up and shutdown of each production run.
- Organizations shall evaluate and document the applicability of this exception for each material, part, or assembly under consideration based upon the production schedule for all customers.
- The evaluation document shall be maintained as an organization-controlled record (7.5.3.2.1); reviewed annually and updated as required. Organizations shall use *CQI-8: Layered Process Audits Guideline, 2nd Edition* to establish an LPA program. Special process assessments organizations shall evaluate the effectiveness of each of the applicable special processes listed below with the associated AIAG manual:
  - Heat Treating—*CQI-9 Special Process: Heat Treat System Assessment, 3rd Edition*
  - Plating—*CQI-11 Special Process: Plating System Assessment*
  - Coating–*CQI-12 Special Process: Coating System Assessment*
  - Welding–*CQI-15 Special Process: Welding System Assessment*
  - Soldering–*CQI-17 Special Process: Soldering System Assessment*
  - Molding–*CQI-23: Special Process: Molding System Assessment*

- Evaluation of implementation effectiveness shall be based on evidence that the organization has a process in place that includes elements such as:
  - Auditors identified
  - Schedule for self-assessment in place (including evidence of schedule adherence)
  - Monitoring of progress
  - Defined corrective action process
  - Organization-controlled record keeping (7.5.3.2.1)
  - Supplier development process (8.4.2.5) identified for applicable suppliers to the organization. Pursuant to IATF 16949 clause 8.4.3.1, this requirement shall also apply to suppliers to the organization who employ the above-listed special processes. Organizations shall evaluate their manufacturing processes, and the manufacturing processes of their suppliers, to establish and document the scope of applicability of this requirement. This document is an organization-controlled record (7.5.3.2.1). Evaluation shall be by self-assessment. The self-assessment shall be conducted annually but may be repeated as needed. The self-assessment may be conducted as part of the organization's internal quality audit or conducted separately. Assessment by a competent second-party auditor (7.2.4) will satisfy the self-assessment requirement for suppliers to the organization.

9.2.2.4 Product audit

Continuing conformance inspection and tests shall be performed during the model year to assure production items or products continue to meet specified requirements and tolerances unless waived in writing by the FCA US release engineer. Any such waiver shall be subject to annual review and renewal. (Refer to PF-8500 and the Product Assurance Testing manual).

9.3–9.3.2.1 NO CHANGES REQUIRED

9.3.3.1 Management review outputs—supplemental

Output from CSRs to the following sections shall provide management review input:

- Design and development planning—Supplemental (8.3.2.1)
- Supplier quality management system development (8.4.2.3)
- Customer satisfaction—Supplemental (9.1.2.1), except as noted earlier

- Quality management system audit (9.2.2.2)
- Manufacturing process audit (9.2.2.3)

Output from Automotive Warranty Management (10.2.5) shall be included in the management review of actual and potential field-failures and their impact upon quality, safety, or the environment.

10–10.1 NO CHANGES REQUIRED

10.2 NONCONFORMITY AND CORRECTIVE ACTION

A written corrective action plan using the 8-Step CA Plan Form shall be submitted to the FCA US supplier quality engineer, as requested, for those issues not already included in the online GIM system.

10.2.3–10.2.4 No changes required

10.2.5 Warranty management systems

Automotive warranty management (AWM). Organizations providing production and non-exempt service parts and components to FCA US shall support improvement in customer satisfaction through pursuit and achievement of warranty reduction targets established by FCA US, where applicable. This shall be accomplished by active participation in the supplier associated warranty reduction program (SAWRP). Organizations shall use CQI-14: Automotive Warranty Management, 3rd ed. to integrate warranty into their quality management system. Evaluation of integration effectiveness shall be based on evidence that the organization has a process in place that includes elements such as:

- Internal auditors identified.
- An established schedule for self-assessment (including evidence of schedule adherence). (A common cycle is every 12 months.)
- A defined continuous improvement process (including evidence of goal-setting and performance evaluation).
- A defined CA process (including evidence of actions taken and verification of effectiveness).
- Organization-controlled record keeping (7.5.3.2.1).
- Progress monitoring (including monthly evaluation of organization's performance to warranty reduction targets established by FCA US).
- A supplier development process (8.4.2.5) identified for applicable suppliers to the organization.

Note: When organizations manage warranty at a corporate level, individual organization sites requiring evidence of compliance to this requirement may reference CQI-14 compliant corporate processes as they pertain to the products and processes at their sites. Evaluation shall be by self-assessment. The self-assessment shall be conducted annually but may be repeated as needed. The self-assessment may be conducted as part of the organization's internal quality audit or conducted separately. The self-assessment shall be conducted using the self-assessment spreadsheet tool from CQI-14. The completed spreadsheet shall serve as a record of the self-assessment. Implementation of AWM shall proceed in three stages:

1. Organization identifies and implements necessary changes to quality management system processes, trains responsible personnel and conducts initial, "baseline" self-assessment.
2. Organization establishes internal performance goals, develops prioritized corrective action plan to achieve these goals and prepares an assessment schedule.
3. Organization monitors performance, continues with self-assessments and updates corrective action plan as required to meet FCA US requirements and internal improvement goals or maintain goal-level performance. Implementation timing for organizations (either new suppliers or current suppliers to FCA US) is summarized in Table 7.6.

AWM Exemptions. The following temporary exceptions apply to organizations that would otherwise be required to implement AWM:

a. *Emergency assumption of business*—Organizations who assume production of parts or components at FCA US's request under emergency conditions are exempt from AWM requirements for six months for these parts or components. The "new supplier/existing program" requirements (above) shall apply thereafter.
b. *Financially distressed suppliers*—Organizations that have been identified by FCA US supplier relations as being financially distressed may, with FCA US supplier quality senior management approval, suspend AWM actions. Such action is considered temporary and will be subject to periodic review by FCA US supplier quality and FCA US supplier relations.

Table 7.6 Implementation timing for automotive warranty management (AWM) requirements

Requirements organization's relationship to FCA US existing vehicle program new vehicle program	Existing Vehicle Program	New Vehicle Program
New supplier	Complete implementation through Stage 2 within six months of award of business. Implementation through Stage 3 to follow within six months of start of production	Complete implementation through Stage 2 before Commercial Launch. Implementation through Stage 3 to follow within six months of Commercial Launch.
Current supplier	Full implementation through Stage 3 required.	Follow timing for "New Supplier/New Vehicle Program" (above) for new parts or components

c. *AWM exemptions*—Organizations that have been identified by FCA US purchasing and supplier quality management as exempt from ISO/TS 16949 or IATF 16949 registration are also exempt from FCA US AWM requirements. However, Mopar® parts or components installed on production vehicles at an assembly plant, a Mopar® custom shop or a dealership at time of sale are considered "production" parts and subject to AWM requirements regardless of the organization's certification status.

Implementation is not required of organizations producing modular assemblies or other products that cannot have warrantable repair assigned to their activity. Implementation is not required of organizations producing parts or components in commodity groups with historically low warranty levels.A list of these low warranty commodity groups is available from the FCA US web page "Supplier Warranty Management—WIS, EWT, GCS, QNA," available in eSupplierConnect.

Organizations whose volume of parts or components supplied in a specific commodity is of low significance may be exempted from FCA US AWM requirements for that commodity. The determination of exemption eligibility for a specific organization–commodity combination is the responsibility of the FCA US supplier quality warranty group.

Note: Questions concerning the program eligibility of individual organizations or commodity groups should be directed to the FCA US supplier quality warranty group at sqwarr@fcagroup.com.

10.2.6 Customer complaints and field failure test analysis

Returned parts analysis—Organizations that provide production or non-exempt service parts or components shall participate in the review, testing and analysis of returned components in accordance with PS-11346 and shall include analysis of the interaction of embedded software, if applicable.

Technical support—Organizations that provide production and non-exempt service parts and components shall provide all necessary support to FCA US in the investigation and resolution of supplier-associated warranty issues.

10.3–10.3.1 NO CHANGES REQUIRED

Sources

FCA US LLC. (2016a). *Customer-Specific Requirements for PPAP*. 4th ed. and Service PPAP, 1st ed. http://www.iatfglobaloversight.org/wp/wp-content/uploads/2016/12/FCA-US-LLC-CSR-PPAP-20161017.pdf. Retrieved on May 10, 2018.

FCA US LLC. (October 2016b). *Customer-Specific Requirements for IATF 16949:2016*. http://www.iatfglobaloversight.org/wp/wp-content/uploads/2016/12/FCA-US-LLC-CSR-IATF-16949-20161017B.pdf. Retrieved on May 10, 2018.

MAQMSR. (September 2017). *Minimum Automotive Quality Management System Requirements for Sub-Tier Suppliers* Minimum Automotive Quality Management (MAQM) requirements for sub tier suppliers. 2nd ed. Rev. AIAG, ANFIA, FIEV, SMMT, VDA—2016.

Note: All Internet sites and procedures referenced in this document may be found in the FCA Covisint system. In order to get into the system, one must have an ID and be authorized by FCA US. All the standards mentioned here may be found on Internet.

General Motor (GM) Customer-Specific Requirements for IATF 16949

Generally, the GM fundamentals of the specific requirements follow the rationale to

1. utilize a layered audit process that includes identifying frequency, schedule, findings, and CA.
2. analyze the risk for all operations using PFMEA and PFMEA methodology.
3. implement a fast response, problem solving process with daily monitoring.

The actual IATF 16949 requirements are:

7.5.3.2 No changes required

7.5.3.2.1 Record retention

The organization's business records shall be retained as specified in GMW15920.

7.5.3.2.2–8.2.3.1.1 No changes required

8.2.3.1.2 Customer-designated special characteristics

The organization shall follow General Motors Key Characteristic Designation System Process GMW15049. Key characteristics shall be applied as per IATF16949:2016 8.3.3.3 Special Characteristics.

8.2.3.1.3–8.3.3 No changes required

8.3.3.1 Product design input

All operations shall be analyzed for risk using a PFMEA. Product requirements shall be identified and failure modes comprehended in the PFMEA. Risk priority number (RPN) values shall be consistently applied using severity, occurrence, and detection ranking tables. Severity shall be based on all risks such as organization risk, customer risk, and end user risk.

8.3.3.2 No changes required

8.3.3.3—8.3.4.3 Special characteristics

The organization shall have a process to identify critical operations within their manufacturing process.

8.3.4.4 Product approval process

The organization shall comply with the AIAG PPAP manual and GM 1927-03 Quality SOR to meet this requirement.

8.3.5—8.3.5.1 No changes required

8.3.5.2 Manufacturing process design output

The organization shall have a method to identify, control, and monitor the high-risk items on those critical operations. There shall be rapid feedback and feed forward between inspection stations and manufacturing, between departments, and between shifts.

8.3.6 No changes required

8.3.6.1 Design and development changes—supplemental

All design changes, including those proposed by the organization, shall have written approval by the authorized customer representative, or a waiver of such approval, prior to production implementation. See also AIAG *Production Part Approval Process* (*PPAP* 4th ed.) manual.

8.4–8.4.2.2 NO CHANGES REQUIRED

8.4.2.3 Supplier quality management system development

This clause applies to suppliers of the organization who are providers of: (a) production materials; (b) production, service, and accessory parts; or (c) heat treating, plating, painting or other finishing services. This clause *does not* apply to indirect or providers of services that add no manufacturing value, which include, but is not limited to distributers, logistics, sequencers, parts packagers, tooling and equipment.

8.4.2.3.1–8.4.2.4 *No changes required*

8.4.2.4.1 Second-party audits

Second-party auditors performing QMS audits must meet the requirements in clause 7.2.4 Second-party auditor compliance in IATF16949:2016, plus meet these additional requirements:

1. The organization (second party) must be IATF16949:2016 certified and not on probation or suspension.
2. The organization (second party) must utilize a qualified ISO lead auditor, or a qualified internal auditor with evidence of their successful completion of training, and a minimum of five internal ISO/TS16949:2009 and/or IATF16949:2016 audits under the supervision of a qualified lead auditor.
3. The organization (second party) must audit annually each qualifying supplier and maintain records of the audit. Qualifying supplier is defined by those suppliers determined to need second-party auditing per clause 8.4.2.4.1 Second-Party Audits. The

duration of these audits must conform to the full application of the audit day requirements table of the current edition of Automotive Certification Scheme for IATF16949 Rules for Achieving and Maintaining IATF Recognition. The second-party audits shall identify an acceptable passing level and include a scoring or ranking to determine which suppliers have passed. The organization shall have documented evidence that they review and follow up on all non-conformances identified in the second-party audit with the intent to close these non-conformances.

8.4.2.5 Supplier development

When a supplier to an organization is so small as to not have adequate resources to develop a system according to IATF16949:2016 or ISO 9001:2015, certain specified elements may be waived by the organization. The organization shall have decision criteria for determining "specially designated small suppliers." Such decision criteria shall be in writing and applied consistently in the application of this provision. The existence and use of such decision criteria shall be verified by third-party auditors.

> **Note 1:** ISO9001:2015 and IATF16949:2016 contain fundamental QMS requirements of value to any size of provider of production materials, production, service, and accessory parts, or heat treating, plating, painting or other finishing services. There are a number of methods to implement a compliant system, so it is recognized that a simpler QMS approach could be used for the smaller suppliers of organizations to which IATF16949:2016 clause 8.4.2.3 applies.

> **Note 2:** "Small" may also refer to volume supplied to automotive.

8.4.3–8.5.1 Information for external providers

8.5.1.1 Control plan

General Motors does not provide waivers to organizations for CP approval because General Motors' signatures on the CP are not required. The organization shall provide measurement, test, and inspection data demonstrating that CP requirements, sample sizes, and frequencies are being met when requested. Sample sizes and frequencies shall be determined based on risk and occurrence of failure modes, and to ensure that the customer is adequately protected from receiving the product represented by the inspection/tests before the results of the inspection/tests are known.

8.5.1.2 Standardized work—operator instructions and visual standards

Standardized work should include what, how, and why tasks are performed. All standardized work shall be followed. Visual standards throughout the facility shall be common, including between facilities building the same platform or product for global quality. Visual standards shall be clearly communicated to all team members that are affected and referenced in the standardized work. Visual standards that differentiate "good" from "bad" shall satisfy customer requirements and be controlled.

8.5.1.3–8.5.1.5 No changes required

8.5.1.6 Management of production tooling and manufacturing, test, inspection tooling and equipment

Where warehouses or distribution centers (distributors) are remote sites, the requirements for management of production tooling may not be applicable.

8.5.1.7–8.5.6 No changes required

8.5.6.1 Control of changes—supplemental

The documented process shall require consideration of a production trial run for every product and process change. Results of the trial run shall be documented.

8.5.6.1.1 Temporary change of process controls

The organization shall have a process for both bypass and deviation. The alternative actions identified on the bypass list shall be customer approved and shall be reviewed using the methodology of the PFMEA to identify the risk. This review shall be documented.

8.6–8.6.1 NO CHANGES REQUIRED

8.6.2 Layout inspection and functional testing

Unless specified otherwise by a GM procuring division, there is no customer-established frequency for layout inspection after receiving PPAP.

8.6.3–9.1.1 No changes required

9.1.1.1 Monitoring and measurement of manufacturing processes

The organization shall have a method for the employee to call or notify for help when an abnormal condition on the equipment or product occurs. A method to call or notify shall be available in all operational areas of the organization. Sufficient alarm limits shall be established for escalation of abnormal conditions and shall match the reaction plan identified in the product's CP.

9.1.1.2 No changes required

9.1.2.1 Customer satisfaction—Supplemental. (New Business Hold)

The certification body of record to the organization shall take the decision to place the organization on immediate suspension* upon receiving notice of GM New Business Hold–Quality.

1. In the event of certification suspension as a result of an organization receiving notice of General Motors New Business Hold–Quality, the organization shall complete a corrective action plan. The organization shall submit the corrective action plan to the certification body of record and to the affected customer(s) within 10 business days of the date of the letter of notification of probation.
   The CA plan of the organization shall be consistent with the affected customer(s) requirements including correction steps, responsibilities, timing information, and key metrics to identify effectiveness of the action plan.
2. Before any suspension can be lifted, the certification body of record shall take the decision to conduct an on-site assessment of appropriate length to verify effective implementation of all CA.
   If suspension is not lifted within four months of its issuance, the certification body of record shall revoke the IATF16949 certificate of the organization. Exceptions to this revocation shall be justified in writing by the certification body based upon its on-site review of the effectiveness of the organization's CA plan and agreement obtained in writing from the authorized GM customer representative.

Note 1: The permitted suspension period for General Motors Europe (GME) is six (6) months.

Note 2: When an organization is placed in NBH after a recertification site audit but before the certificate for recertification is issued: (a) the certification body shall issue the certificate in accord with the IATF *Rules* and (b) the certification body shall then place the new certificate in immediate suspension with the rules for lifting such suspension appropriately applied.

* See *Automotive Certification Scheme for IATF16949, Rules for Achieving and Maintaining IATF Recognition.*

BIQS certification

Organizations shall achieve and maintain built in quality supply base (BIQS) or quality supply base (QSB) certification. The organization whose BIQS or QSB certification is revoked (withdrawn) or expired shall notify its certification body within five (5) business days after revocation or expiration. Lack of the organization having a BIQS or QSB certification shall result in a major finding by the organization's certification body.

For a certification body to close this major finding during the verification audit, the organization shall have either (1) achieved BIQS certification or (2) a documented action plan, confirmed by the GM SQE or SQE designee, detailing the steps, improvements, with target dates, being made to achieve BIQS certification.

Note 1: Exclusions to BIQS certification are shown as WAVD (waived) or NAN (No Audit Needed).

Note 2: The GM system will indicate the designation of WDRN (Withdrawn) for expired or revoked status.

CSII (controlled shipping–level 2)

The organization shall notify its certification body within five (5) business days after being placed in Controlled Shipping–Level 2 (CSII) Status. For CSII activities that are open during an audit, the organization's certification body shall verify that an effective CA process is underway and, if closed, that the CA have been implemented and read across to the entire organization's site for similar processes and/or products. The organization's certification body shall also investigate any CSII activities that have occurred and are closed between surveillance audits.

Note: The GM special status condition of CSII is a performance indicator of an organization's product realization problems. Such status should have resolution, or credible resolution and corrective plans in place, confirmed by the customer.

Process specific audits

The organization shall audit specific manufacturing processes annually to determine their effectiveness. Applicability and effectiveness of these processes shall be determined utilizing the most current version CQI standard (see Table 7.7). The effectiveness evaluation shall include the organization's self-assessment, actions taken, and that records are maintained.

Table 7.7 Audit of specific processes and associated standards

Process	Standard
Heat treating processes	CQI-9 Heat Treat System Assessment
Plating processes	CQI-11 Plating System Assessment
Coating processes	CQI-12 Coating System Assessment
Plastics molding processes	CQI-23 Molding System Assessment
Solder processes	CQI-17 Soldering System Assessment
Casting processes	CQI-27 Casting System Assessment

Note 1: Second-party assessment must be performed by a competent auditor. An auditor is competent if they meet the following requirements: (a) they shall be a qualified ISO lead auditor, or a qualified internal auditor with evidence of their successful completion of training, and a minimum of five internal ISO/TS16949:2009 and/or IATF16949:2016 audits under the supervision of a qualified lead auditor; and (b) they shall have a minimum of five (5) years' experience working with the process that is being audited or a combination of experience and education in the specific process.

Note 2: Audit findings must be addressed in an action plan, with champion(s) assigned and reasonable closure dates

Sources

General Motors. (2016). *IATF 16949: Customer Specific Requirements*. General Motors Company.

General Motors. (November 28, 2017). *IATF Guidance on the Application of Customer Specific Requirements (CSR) and Supplier Codes.*

General Motors. (December 2017). *IATF Oversight Certification Body Communiqué: CB COMMUNIQUE # 2017-012.*

section two

Selected specific issues concerning APQP

chapter eight

Teams

Definition

We all know that "QUALITY" is a team discipline. That means everyone is responsible for quality. However, the "quality discipline" as practiced has deviated from that principle, and we view it as a separate administrative function of the organization. It is unfortunate, but that is where we are.

When we practice the methodologies of APQP, PPAP, FMEA, and many other activities the "team" is the center of that activity. The reason for this is that *no* individual can perform *all* the requirements needed for their completion. Therefore, the "need for the team" is essential for completion of the tasks based on experience and knowledge required. So, to appreciate what the team can do, let us decompose the word as an acronym.

T – together
E – everyone
A – accomplishes
M – more

One can see when the word "team" when is thus decomposed, it shows that cooperation from everyone is necessary for success. This cooperation is called *synergy*.

Therefore, the purpose of creating teams is to provide a framework that will increase the ability of employees to participate in planning, problem-solving, and decision-making to better serve customers. After all, increased participation promotes:

- A better understanding of decisions
- More support for and participation in implementation plans
- Increased contribution to problem-solving and decision making
- More ownership of decisions, processes, and changes

In order for teams to fulfill their intended role of improving organizational effectiveness, it is critical that teams develop into working units focused on their goal, mission, or reason for existing. They do this by effectively progressing and understanding the key ingredients through the stages of team development. So, let us focus then on how we get that *success* through the "team" concept.

- The first and perhaps the most important thing is to identify and define the *purpose* of the team. In its simplest terms, a team is an interdependent group of employees *who unite around a particular task, project, or objective.* That unification *must* be in place before anything else is undertaken.
- The second important ingredient that will solidify the existence of a successful team is the appropriate *culture*: if anyone truly values and wants to encourage teamwork and collaboration, your organization's culture (management, union, and employees at large) must support the activities and provide appropriate skills to do these activities as needed. It is imperative that management must give the authority and responsibility to the team so that the necessary actions may be taken care of. Therefore, management needs to demonstrate leadership and create a work environment that expects, fosters, rewards, and recognizes teamwork.

  It is management leadership style that facilitates cooperation and adjustment to culture. One of the fundamentals is to be serious enough to select an effective team with excellent interpersonal communication dynamics and relationships. To this end, the selected members *must* be clear about the purpose for the team and about each other's roles on the team. Further, the team must have the appropriate knowledge (theoretical and process) to figure out how to evaluate the progress and the effectiveness of the team in relation to the problem at hand.
- The third ingredient is *empowerment* given to the team members. That means that the team must have the *authority* and *responsibility* to take actions regarding the pending solutions of the problem. *Fear of intimidation in all forms must be absent.* Here we must differentiate two terms within the concept of empowerment. They are: (a) commitment, which is primarily a function of management and must be demonstrated in terms of continuous encouragement (delegation) and providing the appropriate and applicable resources as needed to employees, and (b) involvement, on the other hand, is primarily an employee function. It is an essential ingredient to improvement as long as the employees "feel" that their suggestions and actions have a real or significant impact on the decisions and actions that affect their jobs. With empowerment, authority, responsibility, commitment and involvement, it must be understood that *none* of these are considered "goals" or specific "tools." Rather they are philosophies that drive and enable management leadership to pursue culture change in the name of "quality" with teams for improvement.

- The fourth ingredient is *communication*. This means that communication should always be two-way, which implies that the team *must* communicate *good* and *bad* issues to management periodically (or as they occur) and **not** wait until the completion of the project. It also means that management must communicate **all** pertinent information to the team as soon as it is known. If everyone believes and practices honest and truthful communication practices then a productive teamwork environment will be the result.
- The fifth and final ingredient is *team success*. Team success is not "a given" at any point. Rather it must be planned, organized, and cultivated by management. Management's influence may be demonstrated in many ways. Perhaps, one of the primary methods of influence is to allow time early on to convert the selected group into an actionable team. That process was developed in 1965 by Dr. B. Tuckman, who identified the four stages of team development as forming, storming, norming, and performing. In the 1977s he added one more stage: adjourning. It is beyond the scope of book to elaborate on each of the stages, although a very short synopsis of each step follows.
  - **Forming:** The first stage is where a group of people comes together to accomplish a shared purpose. More often than not this stage is used for introductions and sharing background knowledge. Their initial success will depend on their familiarity with each other's work style, their experience on prior teams, and the clarity of their assigned mission. The role of the **sponsor (champion)** is to break the ice of the initial meeting and help the group members get to know each other through team-building activities or just a listening ear.
  - **Storming:** This is the stage for active challenging. It is a struggle between members to make sure that everyone is on the same level of understanding about roles and expectations. Quite often in this stage disagreements about mission, vision, and ways to approach the problem or assignment are a reality. This disagreement sometimes becomes a struggle and very vocal primarily due to combined issues of misunderstanding in the directives about the intent of the team, the unfamiliarity of the individual participants, and the luck of knowledge as to how to interact and communicate with others. The role of the **sponsor (champion)** in this stage is continuously that of a helper. His or her services may be needed to help the group members familiarize themselves with each other by offering team-building activities or just listening to their concerns. In addition, the sponsor may need to help the leader clarify each of the assignments and expectations so that success may prevail.

- **Norming:** In this stage we observe the first indications of team formation. We see the members consciously or unconsciously form working relationships that enable progress on the team's objectives. We observe significant agreements about their own roles and a functional improvement towards working together for the pre-set objectives. The role of the **sponsor (champion)** in this stage is to audit the team for any updates or specific needs that may be needed and making sure that the goal of the team is on schedule.
- **Performing:** This is the stage where the efficiency of the team is apparent as relationships between team members have solidified and team processes are understood more often than not unconsciously. In this stage the team has come to be a functional team as real work is progressing. The role of the **sponsor (champion)** in this stage is to ask for periodic updates from the team. Help solve problems and provide input as needed. Make sure that team members are communicating with all of the other appropriate parties in the workplace. The sponsor makes sure that (a) the team *is not* operating in a vacuum and (b) breaks up or facilitates bottlenecks.
- **Transforming:** This is the stage during which team performance is going so well that members exhibit enthusiasm, satisfaction with their experience, and pride in their accomplishment. The role of the **sponsor (champion)** here is to make sure that everything is continuing as well as expected; indeed, the role of the sponsor is very minimal if it is needed at all.
- **Ending (Adjourning):** This is the stage where the team has completed its mission or purpose and it is time for team members to pursue other goals or projects. In adjourning, the role of the **sponsor (champion)** is to make sure that the team schedules an ending ceremony. The content should cover: (a) a general debrief; (b) a discussion as to whether the team could have been more successful, and if so, what could have been done; (c) an open-minded constructive criticism of the process; (d) recognition of the team and exemplary individual efforts; and (e) celebrate.

Not every team moves through these stages in order, and various activities—such as adding a new team member—can send the team back to an earlier stage while the new member is incorporated. We must remember that in all cases team members (a) operate with a high degree of interdependence, (b) share authority and responsibility for self-management, (c) are accountable for the collective performance, and (d) work toward a common goal and shared rewards(s). After all, a team becomes more than just a collection of people (group) when a strong sense of mutual commitment creates synergy, thus generating performance greater than the sum of the performance of its individual members.

The length of time necessary for progressing through these stages depends on the experience of the members, the support the team receives, and the knowledge and skill of the team members. It must emphasized that the stages mentioned apply to teams that are not expected to stay formed forever. In the case of a department team, a social media team, a customer service team, and so forth, the same stages apply to these ongoing teams except the ending doesn't occur. In any case, we can say that the development of a successful team and teamwork culture is certainly difficult and exciting.

Successful team that last

Hundreds of books and articles have been written about teams, their success and sustainability (See, e.g., Stamatis, 2014, 2015; Llopis, 2012; Heathfield, 2018; Tuckman, 1965; Tuckman et al. 1977; Bales, 1965; Brown, 1999; Forsyth, 1990, 1998, and many others). Here we will summarize some of the most common practices in the literature that build successful teams that last.

We begin with the responsibility of leadership. It takes great leadership to build great teams, leaders who are not afraid to change course, make difficult decisions, and establish standards of performance that are constantly being met—and improving at all times. Whether in the workplace, professional sports, or your local community, team building requires a keen understanding of people, their strengths, and what gets them excited to work with others. Team building requires the *management of egos* and their constant demands for attention and recognition—not always warranted. Team building is both an art and a science, and the leaders who can consistently build high-performance teams are worth their weight in gold. For a short discussion on leadership as it relates to IATF 16949 see appendix A. For a lengthy discussion of teams as they relate to APQP see Stamatis, 1998 pp. 33–50; Stamatis, 2016 pp. 37–42, 343–344, and 372–374.

Building organizations requires the know-how to build long-lasting teams. This is why most managers never become leaders, and why most leaders never reach the highest pinnacle of leadership success. It requires the ability to master the "art of people" and knowing how to maneuver hundreds (if not thousands) of people at the right place and at the right time. It means knowing how each person thinks and how to best utilize their competencies correctly at all times. So, what is the leader's responsibility for success? The following six items seem to be the cornerstones that a leader must practice for excellent team formation (Stamatis, 1997, 2014, 2015; www.smartsheet.com, ND, Llopis, 2012; Heathfield, 2018; Tuckman, 1965; Tuckman et al. 1977; Bales, 1965; Brown, 1999; Forsyth, 1990, 1998).

1. **Be aware of how you work and how you are perceived:** As the leader of the team, one must be extremely aware of one's leadership style and techniques. Are they as effective as you think? How well are they accepted by the team you are attempting to lead? Evaluate yourself and be critical about where you can improve, especially in areas that will benefit those whom you are leading. Though you may be in charge, how you work may not be appreciated by those who work for you. This implies that a leader must *make others feel safe to speak-up: they have to understand that just by their title alone may intimidate.* Successful leaders deflect attention away from themselves and encourage others to voice their opinions. Never *belittle* or *embarrass* anyone in front of others!
2. **Get to know the rest of the team:** Naturally, leaders must be accountable for their actions, but just as important is to make the time to get to know the team and encourage camaraderie. Leaders must communicate with *words and actions* that the team is important and the team's function matters for the improvement of the organization. The leader must communicate, facilitate, guide, and coach based on the knowledge of the members. If the leader does not take the time to learn about the members, he or she projects indifference. The more one knows about the team members, the more effective his or her leadership becomes to match unique areas of subject matter expertise and or competencies to solve problems and seek new solutions.
3. **Clearly define roles and responsibilities:** Generally, the roles of the team composition are set in the first meeting of the group based on the consensus of the group. However, the selection of the champion, leader, and the participants are the role of the management leader—usually the champion.
4. **Be proactive with feedback:** Feedback is the key to assuring any team is staying on track, but more importantly that it is improving each day. Feedback should be proactive and constant. Many leaders are prone to wait until a problem occurs before they give feedback. Because teams are different the feedback can be both formal and informal. However, keep in mind that too much formality may not communicate authenticity and may not have the expected impact.
5. **Acknowledge and reward:** Profoundly important is the notion of acknowledgement and reward. Everybody loves recognition, but is most appreciative of respect. Take the time to give team members the proper accolades they have earned and deserve. (You may consider both individual recognition for exemplary performance and recognition for a combined team effort at the completion of the project.) For the acknowledgement and reward to be effective it must be genuine and timely.

6. **Always celebrate success:** When the team is finished with the project, make sure you celebrate the success. This celebration goes beyond acknowledgment—this is about taking a step back and reflecting on what you have accomplished and what you have learned throughout the journey.

The underlying success characteristic of the above six items is the issue of leadership. Leadership may or may not be management. However, all leaders (management and non-management ranks) are only as successful as their teams, and the great ones know that with the right team dynamics, decisions, and diverse personalities, everyone wins in the end. For this to happen however, leaders must: (a) *Make decisions.* This means that their focus is on "making things happen" at all times—decision-making activities that sustain progress. (b) *Communicate expectations.* By doing so, they remind their colleagues of the organization's core values and mission statement—ensuring that their vision is properly translated and actionable objectives are properly executed. Furthermore, in communicating, successful leaders must not focus on protecting their domain—instead they expand it by investing in mutually beneficial relationships. As good relationships develop the probability increases that success will follow. (c) *Challenge people to think.* This means that successful leaders understand their colleagues' mindsets, capabilities, and areas for improvement. They use this knowledge/insight to challenge their teams to think and stretch them to reach for more. Encourage and enable them to think of "outside of the box" solutions. (d) *Be accountable to others.* This means that leaders understand the inherent power of their position and they are confident in being mentors to their colleagues and employees. This implies that their confidence translates into allowing their colleagues to manage them. Make no mistake about it: this does not mean they allow others to control them, but rather become accountable for assuring they are being proactive to their colleagues' needs. Beyond mentoring and sponsoring selected employees, being accountable to others is a sign that the leader is focused more on the team's success than their own. (e) *Lead by example.* This means that excellent leaders lead by example. *Walk the talk*! Successful leaders practice what they preach and are mindful of their actions. They know everyone is watching them and therefore are incredibly intuitive about detecting those who are observing their every move, waiting to detect a performance shortfall. (f) *Measure and reward performance.* This means performance must be evaluated and rewarded by leaders. Successful leaders never take consistent performers for granted and are mindful of rewarding them. (g) *Provide continuous feedback.* This means that feedback—positive or negative—is imperative for continual improvement. Therefore, leaders must provide feedback, and they must welcome reciprocal feedback. However, in order for the feedback to be constructive and not threating, it must be free of any intimidation from either partner. In giving feedback leaders become teachers for improvement in specific ways, depending on the project. A good teacher is a mentor. (h) *Properly allocate and deploy talent*. This means that leaders must know their talent pool and how to use it. If they do not, it is imperative that they ask questions, ask for recommendations so that their choices for team participants and priority projects can be identified appropriately. (i) *Problem solver: avoid procrastination*. This means that leaders must be excellent problem solvers and not procrastinate in addressing issues and problems that plague the organization. That knowledge should be passed on to the teams for appropriate action, even if the action may be uncomfortable based on the *status quo*. Always remember that getting ahead in life is about doing the things that most people don't like doing. That is why successful leaders in all walks of life create a positive energy and attitude around them. They know how to set the tone and bring an attitude that motivates their colleagues to take action. As such, they are likeable, respected, and strong willed. They don't allow failures to disrupt momentum. (j) *Genuinely enjoy responsibilities.* This means that leaders must love to be leaders not for the sake of power but for the meaningful and purposeful impact they can create. Because they serve as the enablers of talent, culture, and results, they must always be enthusiastic, proactive, determined, and committed to "true" improvement in all aspects of the organization. They must be totally committed to sustainability, i.e., leaving the organization in a better state than they found it. For more information on the leadership of top management as required by the IATF 16949 see Appendix A.

Why teams fail

Failures are part of life. However, as disappointing as they are, if used as a learning activity then failures become very important in problem solving and in any other environment where they occur. Team failures are not any different. However, there are some key things that, if managers and leaders pay attention, can be minimized or completely avoided. Here are some suggestions:

- **Managers pay lip service to employee empowerment, but do not really believe in its power.** When you empower employees, they grow their skills and your organization benefits from their empowerment. Employees know when you are serious about employee empowerment and when you understand and *walk your talk*. Half-hearted or unbelievable employee empowerment efforts will fail.
- **Managers don't really understand what employee empowerment means.** Employee empowerment is a philosophy that enables people to make decisions about their job.

- **Managers fail to establish boundaries for employee empowerment.** In your absence, what decisions can be made by staff members? What decisions can employees make day-by-day that they do not need to have permission or oversight to make? These boundaries must be defined or employee empowerment efforts fail.
- **Managers have defined the decision-making authority and boundaries with staff, but then micromanage the work of employees.** This is usually because managers don't trust staff to make good decisions. Staff members know this and either craftily make decisions on their own and hide their results, or they come to you for everything because they don't know what they really can control.
- **Second-guess the decisions of employees you have given the authority to make a decision.** The anathema of empowerment is second-guessing by leaders. Leaders can help employees (subordinates) make good decisions by mentoring, coaching, training, and providing necessary information when appropriate and applicable. Certainly, leaders need to provide growth and challenging opportunities and goals that employees can aim for and achieve. This means teach them how to take over when you move on. After all, *the primary job of any management leader is to train their replacement*. If the managers (or the organizational culture) are not conducive to creating a work environment that helps foster the ability and desire of employees to act in empowered ways, the team effort and empowerment strategy will fail for sure!
- **Managers abdicate all responsibility and accountability for decision-making.** When managers or the organizational culture is eager to punish, ridicule, blame, punish, or intimidate employees for failures, mistakes, and less than optimum results, the employees—no matter what their capacity—will avoid employee empowerment in future project assignments, or they will identify reasons why the failure was the fault of management (and, in some cases, specific leaders for allowing barriers to impede the ability to be successful).
- **When employees feel under-compensated, under-titled for the responsibilities they take on, under-noticed, under-praised, and under-appreciated, don't expect results from employee empowerment.** Unappreciated employees will deliver very unfavorable message to any organization through word of mouth to other employees—no matter what restrictions have been set by leaders. The result of this communication—unofficial as it is—has catastrophic results for both teams and empowerment. Therefore, make sure the responsibilities and accountability match the job that the person(s) is doing, with the job in the job description—or change it. Accurate accountability helps in growth and establishing competent, capable, and successful employees in the organization.

chapter nine

Risk analysis

Overview

Much has been said about risk and the ISO standards over the last five or so years. As much discussion as has been devoted to it, the fact is that the "risk concept" is immature in the ISO documents. Why? Because of the 157 items in the ISO standards the word "risk" is used in 45 unique definitions, of which 21 specifically address hazards. On the other hand, in the IATF 16949 standard, the word "risk" is sprinkled throughout, and it is especially emphasized in reference to prevention actions and FMEAs.

In both ISO and IATF, the definitions are abundant as they consider events with negative outcomes with a very specific language such as: "A function of the probability of occurrence of a given threat and the potential adverse consequences of that threat's occurrence."

To be sure, that is a very unique definition. However, it is not all inclusive. For example, the PMBOK (2000, Ch. 11; 2017, Ch. 11) goes further by including concerns that deal with the effect of uncertainty, the combination of the consequences of an event and the associated likelihood of its occurrence, and uncertain events or conditions that, if they occur, have a positive or negative effect on a project's objectives.

The term "risk," used in the ISO standards and the new IATF standard, "pertains to safety and/or performance requirements in the context of meeting applicable regulatory requirements at minimum." FMEA is a standard technique used to assess and evaluate potential risks in the design development phase, which continues during production process controls. This and other related risk assessment techniques (i.e., fault tree analysis, warranty analysis, SWOT analysis [strengths, weaknesses, opportunities, threats], event tree analysis, business continuity planning, BPEST [business, political, economic, social, technological] analysis, real option modeling, decision taking under conditions of risk and uncertainty, statistical inference, measures of central tendency and dispersion, PESTLE [political, economic, social, technical, legal, environmental], etc.) can also be used to incorporate other aspects of the QMS (Stamatis, 2014). The specific definition per the ISO standards includes three related definitions:

1. *Risk definition*. Risk is the potential of gaining or losing something of value. Values (such as physical health, social status, emotional well-being, or financial wealth) can be gained or lost when taking a risk

resulting from a given action or inaction, foreseen or unforeseen. Risk can also be defined as the intentional interaction with uncertainty [EN ISO 14971:2012, 2.16].

2. *Risk management.* Risk management is the identification, assessment, and prioritization of risks (defined in ISO 31000 2nd ed. 2018 as the effect of uncertainty on objectives) followed by coordinated and economical application of resources to minimize, monitor, and control the probability and/or impact of unfortunate events or to maximize the realization of opportunities [EN ISO 14971:2012, 2.22].
3. *Risk assessment.* Risk assessment is the determination of quantitative or qualitative estimate of risk related to a well-defined situation and a recognized threat (also called hazard). Quantitative risk assessment requires calculations of two components of risk (R): the magnitude of the potential loss (L), and the probability (p) that the loss will occur (Stamatis, 2014).

The need for risk analysis is defined in several clauses of the standard. For example, in ISO 13485:2016, we find that risk is referred to either directly or indirectly throughout the standard, but specifically in sections (clauses) 4.1; 7.4,1; 7.4.2; 8.2; 8.2.1; 8.3; and 8.3.4.

The issue of risk has become a very important quality requirement and is viewed as a preventive tool in any QMS, APQP, and FMEA of any organization. In fact, for most organizations, there is no need to have a separate section to identify "preventive action," as the concept of preventive action is expressed through a "risk-based" approach, in which the QMS requirements are identified and implemented. For example, the ISO 9001:2015 makes risk-based thinking a requirement by specifically calling for *risks and opportunities* to be determined and addressed. This is a very strong statement to demonstrate that risks are to be identified, corrective actions must be planned, and implementation has to be monitored for the specific risks under consideration.

Reading the standards, one concludes without hesitation that their intent as far as risk is concerned is (a) to provide confidence in the organization's ability to consistently provide customers with conforming goods and services and (b) to enhance customer satisfaction.

The concept of "risk" in the context of the international standards relates to the uncertainty in achieving these objectives.

Given that this is the primary concern, the standards articulate the notion of risk-based thinking, which is defined as something we all do automatically and often sub-consciously. The concept of risk has always been implicit in ISO 9001, but the 2015 revision makes it more explicit and builds it into the whole management system. So, risk-based thinking now has become part of the process approach. The intent here is to make sure that risk-based thinking makes preventive action part of the routine process. This is also very true in the APQP process.

To be sure, risk is often thought of only in the negative sense. However, risk-based thinking can also help to identify opportunities. This can be considered the positive side of risk, which up to recent years has not been recognized as such. The result of this thinking is to:

- Improve customer confidence and satisfaction,
- Assure consistency of quality of goods and services,
- Establish a proactive culture of prevention and improvement, and
- Have successful companies intuitively take a risk-based approach.

In ISO 9001:2015

Risk is implied throughout the ISO 9001:2015 and ISO 14001:2015 standards, as it is referenced wherever the words "planning"—as in clause 6.0—"prevention," and "effectiveness" are mentioned. The reason for this general inclusion is because the ISO standards consider risk to be an *inherent characteristic* in all aspects of a QMS.

Therefore, in a given organization, *risk-based thinking* must be prevalent in any management review because it is a vital element in the continual improvement process focused on prevention. Risk-based thinking must be demonstrated with risk register documentation and must be available for review during audits. A risk register is documented information that validates an organization has done risk-based thinking.

It must be emphasized here that risk-based thinking is indeed PA and unequivocally is everybody's business! As a consequence, it must become an integral part of the organizational culture.

It is significant to note that ISO 9001:2015 does not automatically require anyone to carry out a full, formal risk assessment, or to maintain a risk register per ISO 31000 (risk management principles and guidelines, although it will be a useful reference, it is not mandated). On the other hand, the standard requires proper identification of what the risks and opportunities are in your organization, which depend on context. That is why the active participation of management is imperative.

The premise of risk evaluation as a QMS is founded in clause 4.4—"Quality management system and its processes. Here we find that the organization shall establish, implement, maintain and continually improve a quality management system, including the processes needed and their interactions, in accordance with the requirements of this International Standard. The organization shall determine the processes needed for the quality management system and their application throughout the organization and shall determine the risks and opportunities in accordance with the requirements of 6.1, and plan and implement the appropriate actions to address them."

Furthermore, we read in clause 4.6—Planning for the quality management system—and clause 6.1—Actions to address risks and opportunities—(specifically, 4.1 and 6.1.1), that when planning for the QMS, the organization shall consider the issues referred to in 4.1 and the requirements referred to in 4.2 and determine the risks and opportunities that need to be addressed to: (a) give assurance that the quality management system can achieve its intended result(s); (b) prevent, or reduce, undesired effects; and (c) achieve continual improvement.

In Section 5—Management responsibility—and clause 6.0—Planning for the quality management system—we see specific actions. For example, in 6.1 we find actions to address risks and opportunities. In 6.1.1 we find that when planning for the QMS, the organization shall consider the issues referred to in 4.1 and the requirements referred to in 4.2 and determine the risks and opportunities that need to be addressed to: (a) give assurance that the quality management system can achieve its intended result(s); (b) prevent, or reduce, undesired effects; (c) achieve continual improvement. Finally, clause 6.1.2 tell us that the organization shall plan (a) actions to address these risks and opportunities; (b) how to: (1) integrate and implement the actions into its quality management system processes (see 4.4); and (2) evaluate the effectiveness of these actions. Actions taken to address risks and opportunities shall be proportionate to the potential impact on the conformity of products and services.

Of course, to appreciate what the ISO standard requires, is to see the individual clauses in context. The organizational context should be addressed from at least the following questions:

- Analyze and prioritize the risks and opportunities in your organization: what is acceptable?
- What is unacceptable? Plan actions to address the risks: how can I avoid or eliminate the risk? How can I mitigate the risk?
- Implement the plan—take action. Check the effectiveness of the actions—does it work? Learn from experience—continual improvement.

Overall, the ISO standard presents an outline of requirements for risk-based thinking as specific requirements. For example:

- Risks and opportunities are determined and addressed.
- Implement actions to address the specific risk(s) identified.
- No requirement for formal methods of risk management or a documented risk management process.
- QMS is considered a preventive tool. (There is no need to have a separate clause or sub-clause titled preventive action. The concept

of preventive action is expressed through a risk-based approach to formulating QMS requirements.)
- Risk is defined in ISO 9000 as the effect of uncertainty. (Uncertainty is the state—even partial—of deficiency of information related to understanding or knowledge of, an event, its consequence, or likelihood. Remember, that not all processes have the same level of risk in terms of the organization's ability to meet its objectives.) Therefore, the consequences of process, product, service or system nonconformities will not be the same for all organizations.
- Opportunities can arise as a result of a situation favorable to achieving an intended result, which means any opportunity is not the positive side of risk. Rather, an opportunity is a set of circumstances that makes it possible to do something—either from a positive or negative situation. This means that taking an opportunity presents different levels of risk and appropriate planning, definition, and evaluation must take place—the earlier the better.

Goal

Generally speaking the ISO's goal is to make sure that the customer is satisfied with the quality system of the supplier in such a way that it is sufficient to deliver an acceptable product. In other words, it is a *system-oriented standard.*

The goal of the automotive QMS standard is an upgrade of the ISO to the new IATF 16949 standard as defined on page 7 of that standard. The new upgrade provides for continual improvement, emphasizing defect prevention and reduction of variation and waste (in the entire supply chain) in much more detail. In essence, the idea is to *harmonize the automotive standards globally.*

Project

Once the goal has been defined, the next concern is to make sure management understands the parameters of the project from at least four perspectives (Stamatis, 2016, 1998):

1. *Resources*: A project team with defined roles and responsibilities must be in place for adequate budget and, if suppliers are involved, then to what degree their involvement is necessary. Of great importance here is to recognize the workload in resource planning.
2. *Timing*: Project plan must be in place with internal and external milestones and a critical path that is practical and realistic to fulfill all planned resources requirements.

3. *Addresses customer requirements*: Any success depends on whether the customer requirements are met as defined and expected. If a problem is found during the analysis of the risk, then there are two options. The first is to move the study along with the predefined contingency plan(s) and hopefully the problem will be resolved. The second option is to escalate (to higher level management) the issue or problem so that they can remove any roadblocks and thereby the concern(s) will be taken care of.
4. *Change*: Change is inevitable. There is no doubt that sooner or later it will happen. The question then becomes how. First and perhaps the most important element of any change implementation is the recognition of effective implementation. This occurs with management involvement. To this end, the ISO 9001 and the IATF 16949 standards provide us some clues. For example: In any change management situation at least, the following items must be considered, such as: (a) the inclusion of any customer requirements, in any form; (b) approvals of any change, in any form—in writing; (c) inclusion of any changes and approvals from the supply chain, in any form—in writing; and (d) escalation agreements as to when necessary and how they should be handled.

The key new actions in ISO 9001that address risks and opportunities in addition to the ones that have been mentioned are:

- 6.1.1 *Actions to address risks and opportunities*: Planning needs to consider the context (issues) of the organization (4.1) and the interested party expectations (4.2) and address the risks and opportunities to help the QMS achieve "its intended results."
- 6.1.2 *Actions to address risks and opportunities*: Actions need to be developed, integrated, and implemented back into the QMS processes (4.1).
- 9.1.3 *Analysis and evaluation*: Effectiveness of actions on risks and opportunities.
- 9.3 *Management review*: (b) changes in external and internal issues and (e) effectiveness of actions on risks and opportunities.
- 10.1 *(Improvement) General*: "Preventing or reducing undesired effects."
- 10.2 *Nonconformity and corrective action*: "Update risks and opportunities" based on corrective actions.

In IATF 16949

As of September 2018, IATF 16949 is the official automotive requirement replacing the ISO 9001/16949:2015. It has become an official standard and not an engineering specification. In the standard, we find that risk is mentioned either directly or indirectly throughout, but specifically in sections (clauses): 4.4.1.2; 6.1.2.1; 6.1.2.2; 6.1.2.3; 7.2.3; 8.1.2; 8.3.2.1; 8.7.1.1; 8.7.1.2; and 8.7.1.6; it is evident that the risk is recognized and addressed.

IATF 16949 has 13 New and 83 changed clauses/sub-clauses—96 out of 128 sub-clauses are new or modified. Specifically, IATF 16949 supports all the core tools (SPC, MSA, FMEA, APQP, and PPAP) of the AIAG and adds a number of specific risk-related requirements to minimize the chance of failure during new program development and to maximize (not optimize) the chance for realization of planned activities. Key clauses that deal directly with managing and mitigating risk are: 8.1.1; 8.2.2.1; 8.2.3.1.1; 8.2.3.1.3; and 8.6.6. See Appendix B for a complete list of the changes.

The comprehensiveness of the new standard is coming to light with further requirements that deal with additional controls for the management of development projects through the cycle, and eventually concludes with a product approval process. Key clauses that deal with this area are: 8.3.3.1; 8.3.4.2; 8.3.4.3; 8.3.5.1; and 8.3.4.4. A summary of new items and changes to the IATF standard follows.

New to IATF

- 4.3.2 *Customer specific requirements*: Included in the scope of the QMS. IATF interprets this to also mean that they are integrated in the processes of the organization, i.e., QMS.
- 4.4.1.1 *Conformance of products and processes*: A general statement that all products and processes, including outsourced and service parts, need to meet all customer, statutory, and regulatory requirements (ref. 8.4.2.2). The IATF interprets this to mean that the organization takes a proactive approach to assess and address risk and move away from inspection. (It focuses on prevention.)
- 4.4.1.2 *Product safety*: Management of functions/characteristics and FMEAs/CPs for product-safety related products and processes. Includes transfer of requirements down the supply chain and traceability.
- 5.1.1.1 *Corporate responsibility*: Define and implement corporate policies for anti-bribery, employee code of conduct, and ethics escalation policy. A practical approach here is to evaluate policies and requirements based on individual risks at each site and then integrate them into the overall organization's business processes. If and when this integration takes place, it will help the organization with other social responsibility and sustainability.
- 5.1.1.3 *Process owners*: Identify process owners to manage the organization's processes. This was always a requirement but is appearing in the standard for the first time.
- 6.1.2.3 *Expansion of contingency planning.*

- 7.2.4 *Second-party auditor competency*.
- 8.3.2.3 and 8.4.2.3.1 *Embedded software*.
- STRENGTHENED—FMEAs and CP requirements (throughout).
- 8.5.6.1.1 *Temporary change of process controls*: Maintain a list of process controls and alternatives and document sufficiently to remove risk and maintain traceability in case there are issues.
- 10.2.5 *Warranty management*: Implement a warranty management system including No Trouble Found. Use customer-prescribed methods.

Changes

- 6.1.2.1 *Risk analysis*: Analyze lessons learned from product recalls, product audits, field returns and repairs, complaints, scrap and rework, and implement action plans. Effectiveness needs to be analyzed and actions integrated into the QMS.
- 7.1.3.1 *Plant, facility, and equipment planning*: Addition of risk identification and risk mitigation, including manufacturing feasibility assessments such as capacity planning (use customer/quoted rates for evaluation as well) for new products or new operations and for proposed changes to existing operations. This needs to be included in management review.
- *Type and extent of control—supplemental*: Strengthened control of outsourced processes and type and extent of control for verification of conformity and includes need for a documented process. IATF 16949 requires criteria and escalation to be based on supplier performance and assessment of risk.
- 8.7.1.5 *Control of repaired product*: Use risk analysis processes such as FMEA to assess risk in repair process. This also needs to be documented. FMEA and control occurrences must be reviewed and a list generated to determine which of these will cause the organization to redo or improve the FMEAs and CPs.
- 4.4.1.2 Product safety—Design FMEA—Process FMEAs—CP.
- 7.1.3.1 Plant, facility, and equipment planning—CP.
- 7.1.5.1.1 Measurement system analysis—CP.
- 7.2.3 Internal auditor competency—PFMEA—CP.
- 7.2.4 Second-party auditor competency—PFMEA—CP.
- 7.5.3.2.2 Engineering specifications—FMEA—CP.
- 8.3.2.1 Design and—development planning—supplemental—FMEA and CP.
- 8.3.3.3 Special characteristics—FMEA—CP.
- 8.3.4.3 Prototype program—CP.
- 8.3.5.1 Design and development outputs—Supplemental—FMEA.
- 8.3.5.2 Manufacturing process design output—FMEA—CP.
- 8.5.1.1 Control Plan—FMEA—CP.

- 8.5.6.1.1 Temporary change of process controls—FMEA.
- 8.6.1 Release of products and services—supplemental—CP.
- 8.6.2 Layout inspection and functional testing—CP.
- 7.1.4 Control of reworked product—FMEA—CP.
- 8.7.1.5 Control of repaired product—FMEA—CP.
- 9.1.1.1 Monitoring and measurement of manufacturing processes—PFMEA—CP.
- 9.1.1.2 Identification of statistical tools—DFMEA—PFMEA.
- 9.2.2.3 Manufacturing process audit—PFMEA—CP.
- 9.3.2.1 Management—review inputs—supplemental—FMEA.
- 10.2.3 Problem-solving—PFMEA—CP.
- 10.2.4 Error-proofing—PFMEA—CP.
- 10.3.1 Continual improvement—supplemental—FMEA. (For a detailed analysis and discussion on the FMEA, see Stamatis, 2003 and 2015.)

VDA.6 (The new FMEA standards of 2018)

There are no ambiguities about the relationship of FMEA and risk. The two are very closely related. However, for all industries—except automotive—the requirements of *all* FMEAs are the same. That is, they follow the specifications of the SAE: J1739, ISO requirements, and specific requirements—as appropriate—from their customers.

In the automotive industry, under the leadership of a German consortium (*Verband der Automobilindustrie*), a new standard has arrived called the VDA.6. It is supposed to harmonize the automotive standards worldwide. In the United States, the standard has met with mixed feelings, although the AIAG has given the standard the green light for further clarification and discussion among its members before full implementation. However, that is not a mandatory requirement for Ford, GM, and FC, so it is up to them if they want to use it or continue with the traditional FMEA approach.

VDA.6 (2017) is an upcoming standard based on European bureaucracy that addresses *only* the automotive FMEA and risk requirements. In many ways it seems very cumbersome and more subjective that the traditional FMEA. However, the new methodology is already being applied in the European automotive community with some continuing resistance in the United States. The proponents of VDA.6 recognize some major changes in both rationale and approach; however, they feel strongly that the implementation of the new standard will have the following expected benefits:

- Supply German automakers with a good quality product,
- Maintain higher quality standards than even IATF 16949,
- Increase overall quality of your products,
- Reduce warranty costs, and
- Minimize risk issues/concerns and problems.

In any case, we find that risk is mentioned either directly or indirectly throughout the standard. However, in some specific sections (clauses) it is emphasized with greater strength:

Existing FMEAs conducted with an earlier version of the FMEA handbook may remain in their original form for subsequent revisions.

- Optionally, the team may decide to transfer the data to the latest form and update the FMEA in accordance with the latest FMEA procedure in order to take advantage of improvements associated with the latest FMEA procedure.
- FMEAs that will be used as a starting point for new program applications should be converted to comply with the new format.
- However, if the team determines that the new program is considered a minor change to the existing product, they may decide to leave the FMEA in the existing format.
- New projects should follow this FMEA procedure if not otherwise defined, unless company procedure defines a different approach.

Flow

- DFMEA contains information that is useful for PFMEA
- Failure causes related to piece-to-piece
- End user failure effects and severity for the failure causes related to product characteristics
- PFMEA contains information that needs alignment with the DFMEA
- Failure effects and severity for failure modes also shown in the DFMEA
- Not all failures causes in a DFMEA are failure modes in a PFMEA

Confidential design and or process

- The sharing of intellectual property between suppliers and customers is governed by legal agreements between suppliers and customers and is beyond the scope of this handbook.
- However, unless otherwise required by contractual agreement, for reasons of intellectual property (IP) protection the DFMEAs and PFMEAs prepared by suppliers for standard or "off the shelf" products should generally be considered proprietary information not given to the customers, but may be shown by special arrangement when requested.

Action Priority

High: The team must either identify an appropriate action to improve prevention and/or detection controls or justify and document why current controls are adequate.

Medium: The team should identify appropriate actions to improve prevention and/or detection controls, or, at the discretion of the company, justify and document why controls are adequate.

Low: The team could identify actions to improve prevention or detection controls. It is recommended that potential Severity 9–10 failure effects with action priority high and medium, at a minimum, be reviewed by management including any actions that were taken. This is not the prioritization of high, medium, or low risk. It is the prioritization of the need for actions to reduce risk.

The proposed model of VDA 6 (The six step model)

- Step 1 Scope (problem) definition: AIAG/VDA, 2017, p. 30
- Step 2 Structure analysis: AIAG/VDA, 2017, p. 33
- Step 3 Function analysis: AIAG/VDA, 2017, p. 37
- Step 4 Failure analysis: AIAG/VDA, 2017, p. 44
- Step 5 Risk analysis: AIAG/VDA, 2017, p. 53
- Step 6 Optimization: AIAG/VDA, 2017, p. 66

Specifically, these steps include:

2.1.1 Purpose: The purpose of the DFMEA scope definition is to define what is included and excluded in the FMEA based on the type of analysis being developed, i.e., system, subsystem, or component. The main objectives of DFMEA scope definition are:

- Definition/selection of which aspects of the design are to be included in the analysis,
- Project plan (of DFMEA activities, ref paragraph 1.6),
- Identifying relevant lessons learned and reference materials that should be used to define the scope, and
- Definition of team responsibilities.

2.2.1 Purpose: The purpose of design structure analysis is to identify and breakdown the design into system, subsystem, and component, parts for technical risk analysis. The main objectives of a design structure analysis are:

- Identification of relevant system elements and definition of a system structure;
- Visualization of the scope of analysis;
- Analysis of relationships, interfaces and interaction between defined system elements; and
- Visualization via, e.g., structure trees, block (boundary) diagrams, etc.

2.3.1 Purpose: The purpose of the design function analysis is to ensure that the functions specified by requirements/specifications are appropriately allocated to the system elements. The main objectives of a design function analysis are:

- Associate functions with the relevant system elements;
- Overview of the functionality of the product;
- Describe each function in detail by using parameter diagrams or other methods;
- Allocation of requirements/characteristics to individual functions;
- Visualization via, e.g., function tree, net, function matrix; and
- Cascade of customer (external and internal) functions with associated requirements for the intended use.

The structure provides the basis so that each system element may be individually analyzed with regard to its functions and requirements. For this, comprehensive knowledge of the system *and* the operating conditions *and* environmental conditions of the system are necessary, including, for example heat, cold, dust, splash water, salt, icing, vibrations, electrical failures, etc.

2.4.1 Purpose: The purpose of the design failure analysis is to identify failure causes, modes, and effects, and show their relationships to enable risk assessment. The main objectives of a design failure analysis are:

- Identification of potential failures assigned to functions in structural elements,
- Establishment of the failure chain (effects, modes, causes),
- Visualization of failure relationships (failure nets and/or FMEA Spreadsheet), and
- Collaboration between customer and supplier (failure effects).

2.5.1 Purpose: The purpose of design risk analysis is to estimate risk by evaluating severity, occurrence and detection, and prioritize the need for actions. The main objectives of the design risk analysis are:

- Assignment of prevention controls (existing and/or scheduled);
- Assignment of detection controls (existing and/or scheduled);
- Rating of severity, occurrence and detection for each failure chain;
- Collaboration between customer and supplier (severity); and
- Evaluation of action priority.

2.6.1 Purpose: The purpose of the design optimization is to determine actions to mitigate risk and assess the effectiveness of those actions. The main objectives of a design optimization are:

- Identification of the actions necessary for improvement,
- Assignment of responsibilities and target completion times for action implementation,
- Implementation and documentation of actions taken,
- Configuration of the effectiveness of the implemented actions,
- Re-assessment of risk after actions taken, and
- Continuous improvement of the design.
- Basis for refinement of the product requirements and prevention and or detection controls.

The primary objective of optimization is to develop actions that reduce risk and increase customer satisfaction by improving the design—see Appendix D for more discussion on FM avoidance. In this step, the team reviews the results of the risk analysis and assigns actions to lower the likelihood of occurrence of the failure cause or increase the robustness of the detection control to detect the "failure cause or failure mode." Actions may also be assigned to improve the design but do not necessarily lower the risk assessment rating. Actions represent a commitment to take a specific, measurable, and achievable action, not potential actions that may never be implemented. Actions are not intended to be used for activities that are already planned, as these am documented in the prevention or detection controls and are already considered in the initial risk analysis.

- If the team decides that no further actions are necessary, "None" or "No revision planned" is written in the remarks column to show the risk analysis was completed. The DFMEA should be used to assess technical risks related to continuous improvement of the design. The optimization is most effective in the following order:
  1. Design modifications in order to reduce the occurrence of the failure cause (FC).
  2. Increase the ability to detect the failure cause or failure mode (PC or FM).
- In the case of design modifications, all impacted design elements are evaluated again.
- In the case of concept modifications, all steps of the DFMEA are reviewed for the affected sections. This is necessary because the original analysis is no longer valid since it was based upon a different design concept.

Method or process of conducting any risk assessment

The generic method for a risk analysis of any kind consists of the following elements, performed, more or less, in the following order.

1. Identify, characterize *threats*,
2. Assess the *vulnerability* of critical assets to specific threats,
3. Determine the *risk* (i.e., the expected likelihood and consequences of specific types of attacks on specific assets),
4. Identify ways to reduce those risks, and
5. Prioritize risk reduction measures.

To appreciate these steps, one must understand and evaluate them in context. This context involves:

1. Identifying risk(s) in a selected domain of interest;
2. Planning the remainder of the process;
3. Mapping out the:
   i. Social scope of risk management,
   ii. Identity and objectives of *stakeholders (process owners)*, and
   iii. Basis upon which risks will be evaluated, constraints;
4. Defining a framework (boundary) for the activity and an agenda for identification; and
5. Developing an analysis of risks involved in the process mitigation or solution of risks using available technological, human and organizational resources and/or methods.

General evaluation criteria for DFMEA

Severity: Whereas the evaluation criteria for a DFMEA for industries other than the automotive still follow the AIAG guidelines (or some modification of them), the automotive industry, by adopting the VDA standard, has quite different proposed criteria. However, since they are not officially published, we can say that the severity is evaluated based on what the end user experiences.

Occurrence: Whereas the evaluation criteria for a DFMEA for industries other than the automotive still follow the AIAG guidelines (or some modification of them), the automotive industry, by adopting the VDA standard, has quite different proposed criteria. However, since they are not officially published we can say that occurrence will be evaluated based on (a) estimated frequency, (b) product experience, and (c) prevention control.

Detection: Whereas the evaluation criteria for a DFMEA for industries other than the automotive still follow the AIAG guidelines (or some modification of them), the automotive industry, by adopting the VDA standard, has quite different potential criteria. However, since they are not officially published, we can say that the detection will be based on the activity performed prior to delivery of the design for production.

General evaluation criteria for PFMEA

Severity: Whereas the evaluation criteria for a PFMEA for industries other than the automotive still follow the AIAG guidelines (or some modification of them), the automotive industry, by adopting the VDA standard, has quite different proposed criteria. However, since they are not officially published, we can say that the severity is evaluated based on what the failure effects are rated for manufacturing, assembly, and the end user.

Occurrence: Whereas the evaluation criteria for a PFMEA for industries other than the automotive still follow the AIAG guidelines (or some modification of them), the automotive industry, by adopting the VDA standard, has quite different proposed criteria. However, since they are not officially published, we can say that the occurrence is evaluated based on will be evaluated based on (a) estimated frequency, (b) process experience, and (c) prevention control.

Detection: Whereas the evaluation criteria for a PFMEA for industries other than the automotive still follow the AIAG guidelines (or some modification of them), the automotive industry, by adopting the VDA standard, has quite different proposed criteria. However, since they are not officially published, we can say that the detection will be based on activities prior to shipment of he product, so that nonconformances may be caught.

Monitoring and system response (MSR)

The scope of a supplemental FMEA for MSR may be established in consultation between the customer and supplier. Typical steps are:

1. Scope definition
2. Structure analysis
3. Function analysis
4. Failure analysis
5. Risk analysis
6. Optimization

Action priority (AR) for FMEA-MSR

The AR is a methodology that allows for the classification of the risks guiding the team in their prioritization of the need for action. (Very important: *This is not the prioritization of high, medium, or low risk. It is the prioritization of the need for actions to reduce risk. At a minimum the statement that "No further Action is needed" must be included.*)

Priority High (H): Highest priority for action. The team *must* either identify an appropriate action to improve prevention and/or detection controls or justify and document why current controls are adequate.
Priority Medium (M): Medium priority for action. The team *should* identify appropriate actions to improve prevention and/or detection controls, or, at the discretion of the company, justify and document why controls are adequate.
Priority: Low (L): Low priority for action. The team *could* identify actions to improve prevention or detection controls.

It is recommended that potential severity 9–10 failure effects with AR high and medium, at a minimum, be reviewed by management including any recommended actions that were taken.

FMEA form

Whereas the form used for the FMEA for industries other than the automotive still follow the classic AIAG format, the automotive, by adopting the VDA 6 standard, has modified the form quite extensively—see Figure 9.1. See Appendix C for all the forms used for FMEA.

Principles

Generally, there are 12 principles that guide the way that risk management is integrated and deployed in ISO 9001/16949:2015; in IATF:16949; in VDA.6, and in ISO 31000:2018:

1. Risk management creates and protects value. Resources expended to mitigate risk should be less than the consequence of inaction.
2. Risk management is an integral part of all organizational processes.
3. Risk management is part of decision-making.
4. Risk management explicitly addresses uncertainty and assumptions.
5. Risk management is systematic, structured, and timely.
6. Risk management is based on the best available information.
7. Risk management is tailored to specific tasks.
8. Risk management takes human and cultural factors into account.

DESIGN FAILURE AND EFFECTS ANALYSIS
(DFMEA)

Company Name: ____________
Engineering Location: ____________
Customer Name: ____________
Model / Year / Platform: ____________
FMEA Team: ____________

Subject: ____________
DFMEA Start Date: ____________
DFMEA Revision Date: ____________
FMEA Due Date: ____________

DFMEA ID Number: ____________
Design Responsibility: ____________
Security Classification: ____________
FMEA Intast: ____________

FMEA Tool: ____________

STRUCTURE ANALYSIS			FUNCTION ANALYSIS			FAILURE ANALYSIS				RISK ANALYSIS					
1. System (Item)	2. System Element / Interface	3. Component Element (Item / Interface)	1. Function of System and Requirement or Intended Output	2. Function of System Element and Intended Performance Output	3. Function of Component Element and Requirement or Intended Output or Characteristic	1 Failure Effects (FE)	Severity (S) of FE	2. Failure Mode (FM)	3. Failure Cause (FC)	Current Prevention Control (PC) of FC	Occurrence (O) of FC	Current Detection Control (DC) of FC or FM	Detection (D) of FC/FM	AP	Filter Code (Optional)

OPTIMIZATION										
Prevention Action	Detection Action	Responsible Person	Target Completion Date	Status: [Untouched, Under Consideration, In Progress, Completed, Discarded]	Action Taken with Pointer to Evidence	Completion Date	Severity (S)	Occurrence (O)	Detection (D)	AP

Figure 9.1 VDA FMEA form.

9. Risk management is transparent and inclusive.
10. Risk management is dynamic, iterative and responsive to change.
11. Risk management facilitates continual improvement and enhancement.
12. Risk management demands of the organization, to be continually or periodically reassessing their systems.

Identification

After establishing the context, the next step in the process of managing risk is to identify potential risks. Risks are about events that, when triggered, cause problems or benefits. Hence, risk identification can start with the source of our problems and those of our competitors (benefit) or with the problem itself. It is important to note here that the problem is recognized at the *actionable root cause level*. It is also important to identify the problem at the *escape point* (EP). The EP is the place of the process when the *root cause* could have been caught, but it was not. Unless we find the EP, we have a perpetual problem. We may find out how to fix the problem, but the fixing is going to be continual as the origin of the root cause will not be known.

So, how do we proceed? Depending on the situation there are several options. However, the most commonly used fall into two categories.

1. *Source analysis*: Risk sources may be internal or external to the system that is the target of risk mitigation (I use mitigation instead of management since by its own definition risk deals with factors of decision-making that cannot be managed). Examples of risk sources are: owners of a project, employees of a company, customer requirements, legal requirements or events of nature.
2. *Problem analysis*: Risks are related to identified threats, for example: the threat of losing money, the threat of abuse of confidential information or the threat of human errors, accidents, and casualties. The threats may exist with various entities, most importantly with process owners, customers, and legislative bodies such as the government. When either source or problem is known, the events that a source may trigger or the events that can lead to a problem can be investigated. For example: stakeholders withdrawing during a project may endanger funding of the project; confidential information may be stolen by employees even within a closed network; natural events may make many people fall casualties.

The chosen method of identifying risks may depend on culture, industry practice, and compliance. The identification methods are formed by templates or the development of templates for identifying source, problem, or event. Common risk identification methods are:

- *Objectives-based risk identification*: Organizations and project teams have objectives. In practical terms, any event that may endanger achieving an objective partly or completely is identified as risk.
- *Scenario-based risk identification*: In scenario analysis, different scenarios are created. The scenarios may be the alternative ways to achieve an objective, or an analysis of the interaction of forces in, for example, a market or a system. Any event that triggers an undesired scenario alternative is identified as risk. Scenarios generally are futuristic in nature, with the probability nevertheless low.
- *Taxonomy-based risk identification*: The taxonomy (classification) in taxonomy-based risk identification is a breakdown of possible risk sources. Based on the taxonomy and knowledge of best practices, a questionnaire is compiled. The answers to the questions reveal risks (Carr et al., 1993).
- *Common-risk checking*: In several industries, lists with known risks are available. Each risk in the list can be checked for applicability to a particular situation (Stamatis, 2014; Common Vulnerability and Exposures list, April 16, 2018).
- *Risk charting* (Crockford, 1986, p. 18): This method combines the above approaches by listing resources at risk, threats to those resources, and modifying factors that may increase or decrease the risk and consequences it is desirable to avoid. Creating a matrix under these headings enables a variety of approaches. One can begin with resources and consider the threats they are exposed to and the consequences of each threat. Alternatively, one can start with the threats and examine which resources they would affect, or one can begin with the consequences and determine which combination of threats and resources would be involved to bring them about.

General comments

Everything we do in life has a risk. Generally, that risk is evaluated (assessed) based on two items. They are:

1. The benefit one gets from performing a task, and
2. The negative consequences from either not doing or performing the task.

In any environment there is always more than one risk involved. So, some options in identifying the available risk options are:

1. Design a new business process with adequate built-in risk control and containment measures from the start.
2. Periodically re-assess risks that are accepted in ongoing processes as a normal feature of business operations and modify mitigation measures.

3. Transfer risks to an external agency (e.g., an insurance company).
4. Avoid risks altogether (e.g., by closing down a particular high-risk business area).

It is imperative to be aware that the financial benefits of risk management are less dependent on the use of a given formula, but are more dependent on the frequency and effectiveness of how risk assessment is performed. The assessment is not always in financial terms. It may be safety, government regulations, customer demands, and so on. In case the assessment is a financial one, make sure an appropriate cost-benefit analysis is performed.

In either case there is a formal approach to evaluating "risk." However, the evaluation may take several steps depending on the organization, industry, and or particular task. In general terms, the assessment is based on the following:

- Identify the risk or benefit. Make sure that all (or at least the pertinent risks) are well defined, including the relationships with adjacent or relative risks or tasks.
- Decide who may be benefitted or harmed, and how (look for magnitude). If possible assign probabilities to the tasks. Here you may consider: safety considerations, cost, quality, capability, technical functionality, etc.
- Assess the risks and take action or no action (remember: taking *no action* is, in fact, an action). Must evaluate criticality and rank that criticality from worst to best.
- Make a record of the findings. This is very important. Therefore, make sure the analysis will cover the things gone right (TGR) and things gone wrong (TGW). The ranking from the previous step should help in the prioritization process. Always remember: some things are more important than others. If all problems are important—at the same level—there is no need for any risk analysis. Risk analysis forces priority items to be addressed first.
- Review the risk assessment. Make sure all owners of the task are present for the review. Decide on a mitigation resolution. For example, high critical items should be pursued right away, medium to low critical items should be tracked, monitored and acted upon when they reach a threshold of concern or they impede the task. The mitigation process is very important as most risk analyses fall apart at this stage because of not having a good definition of the risk being evaluated and or the owners of the task did not participate in the analysis. Typical approaches may be:
  - *Assume/accept*: Acknowledge the existence of a particular risk and make a deliberate decision to accept it without engaging in special efforts to control it. Approval of a project or program

leaders is required. Management's involvement in this stage is mandatory. If they are not involved, the risk may be the wrong one. Collaborate with the operational users to create a collective understanding of risks and their implications. Risks can be characterized as impacting traditional cost, schedule, and performance parameters. Risks should also be characterized by their impact on mission performance resulting from reduced technical performance or capability. Develop an understanding of all these impacts. Bringing owners into the mission impact characterization is particularly important to selecting which "assume/accept" option is ultimately chosen. Owners will decide whether accepting the consequences of a risk is acceptable. Provide the owners with options (i.e., the vulnerabilities affecting a risk, countermeasures that can be performed, and residual risk that may occur). Help the owners understand the costs in terms of time and money.

- *Avoid*: Adjust program requirements or constraints to eliminate or reduce the risk. This adjustment could be accommodated by a change in funding, schedule, or technical requirements. Again, work with all owners (stakeholders) to achieve a collective understanding of the implications of risks. Provide all concerned with projections of schedule adjustments needed to reduce risk associated with technology maturity or additional development to improve performance. Identify capabilities that will be delayed and any impacts resulting from dependencies on other efforts. This information better enables users to interpret the operational implications of an "avoid" option.
- *Control*: Implement actions to minimize the impact or likelihood of the risk. Help control risks by performing analyses of various mitigation options. For example, one option is to use a commercially available capability instead of a contractor-developed one. In developing options for controlling risk in your program, seek a partner to conduct a benchmarking study, so that your organization can benefit from other organizations' successes with either similar or identical problems of risk.
- *Transfer*: Reassign organizational accountability, responsibility, and authority to another stakeholder willing to accept the risk. Management's commitment is necessary here. However, keep in mind that reassigning accountability, responsibility, or authority for a risk area to another organization can be a double-edged sword. It may make sense when the risk involves a narrowly specialized area of expertise not normally found in program offices, but transferring a risk to another organization can result in dependencies and loss of control that may have their own complications. Position yourself and your customer to consider a transfer option by acquiring

and maintaining awareness of organizations within your customer space that focus on specialized needs and their solutions. Acquire this awareness as early in the program acquisition cycle as possible when transfer options are more easily implemented.

- *Watch/monitor*: Monitor the environment for changes that affect the nature and/or the impact of the risk. Once a risk has been identified and a plan put in place to manage it, there can be a tendency to adopt a "heads down" attitude, particularly if the execution of the mitigation appears to be operating on "cruise control." Resist that inclination. Periodically revisit the basic assumptions and premises of the risk. Scan the environment to see whether the situation has changed in a way that affects the nature or impact of the risk. The risk may have changed sufficiently so that the current mitigation is ineffective and needs to be scrapped in favor of a different one. On the other hand, the risk may have diminished in a way that allows resources devoted to it to be redirected.
- Design a new business process with adequate built-in risk control and containment measures from the start.
- Periodically re-assess risks that are accepted in ongoing processes as a normal feature of business operations and modify mitigation measures.
- Transfer risks to an external agency (e.g., an insurance company).
- Avoid risks altogether (e.g., by closing down a particular high-risk business area).

Each of these options requires developing a plan that is implemented and monitored for effectiveness.

From a system's engineering perspective, common methods of risk reduction or mitigation with identified program risks include the following, listed in order of increasing seriousness of the risk (ISO 31000:2009, revised with ISO 31000: 2018: Risk management—Principles and guidelines).

1. Intensified technical and management reviews of the engineering process
2. Special oversight of designated component engineering
3. Special analysis and testing of critical design items
4. Rapid prototyping and test feedback
5. Consideration of relieving critical design requirements
6. Initiation of fallback parallel developments

Essentially, the key capabilities of a risk-enabled quality management solution include, but are not limited to, the following:

- **An integrated risk register:** The need for a centralized place to record and monitor individual hazards and risk items. The importance here is to demonstrate consistency. The minimum content for this document are: dates, description of risk, risk type (business, project and stage), severity of effect, likelihood of occurrence (low < 30%, medium 31%–75% or high > 75%), countermeasures, status and any other quantitative value that may be applicable and appropriate. A typical risk register based on the work of Kokcharov (2015) is shown in Table 9.1

- **Flexible risk tools:** The need for a matrix of tools and to be able to activate risk assessment tools to demonstrate the link between audits to risk deviations or regulatory requirements. It is a way to track the effectiveness of the tools as they are applied to specific risks.
- **Risk-based effectiveness checks:** This requirement is used for verification of corrective action and evaluating improvement requirements.

With the proliferation of risk awareness, we must also mention the future of a new wave within quality assurance called Quality 4.0 (Jacob, 2017), which deals with the factory of the future and the industrial Internet of things (IIoT). Quality 4.0 and the digital transformation of quality, combines the elements of technology, process innovation, and risk-based thinking to provide greater visibility and control into quality processes. Risk management tools will be more important than ever under Industry 4.0 (Jacob, 2017), with potential concerns around:

- **Security:** Interconnected cyber-physical systems raise questions about data security and protecting proprietary information.
- **Systems reliability:** Keeping automated systems up and running is a top priority and could severely affect quality if not properly managed.
- **Data talent:** Managing the flow and volume of data that Industry 4.0 is designed to deliver requires skilled data scientists. Unfortunately, the current shortage of data experts is only getting worse, so companies will need to deliberately recruit and cultivate digital talent.

Table 9.1 A typical risk register for a project that includes 4 steps: Identify, analyze, plan response, monitor, and control

Identify		Analyze				Plan response	Monitor and control	
ID	Description	Category	Probability	Estimated impact	Additional workload days	Status	Response cost	Costs

Key items of concentration are: big data databases, predictive maintenance, OEE), risk management, machine learning (with artificial intelligence built into them for self-correction mechanisms, and supply chain interactions.

Best practices and lessons learned

Top management should review the organization's QMS at planned intervals to ensure its continuing suitability, adequacy, and effectiveness. The management review should be planned and carried out taking into consideration: (a) the status of actions from previous management reviews; (b) changes in external and internal issues that are relevant to the quality management system including its strategic direction; (c) information on the quality performance, including trends and indicators for: (1) nonconformities and corrective actions, (2) monitoring and measurement results, (3) audit results, (4) customer satisfaction, (5) issues concerning external providers and other relevant interested parties, (6) adequacy of resources required for maintaining an effective quality management system, and (7) process performance and conformity of products and services; (d) the effectiveness of actions taken to address risks and opportunities (see clause 6.1); and (e) new potential opportunities for continual improvement.

What are the actions one must take to optimize the lessons learned? Fundamentally, the actions should be considered with the following in mind: they must be (a) timely, (b) within reasonable cost, (c) doable within time and specification requirements, and (d) have appropriate resources.

- What actions are needed?
    - Make sure you have the right exit criteria for each action. For example, appropriate decisions, agreements, and actions resulting from a meeting would be required for exit, not merely the fact that the meeting was held. Always remember that if there is no agenda and no meeting notes, *there was no* meeting!
    - Look for evaluation, proof, and validation of met criteria. Consider, for example, metrics or test events. The metrics must be doable.
    - Include any and all owners relevant to the step, action, or decisions.
- When must actions be completed?
    - Backward Planning: evaluate the risk impact and schedule of need for the successful completion of the program and evaluate test events, design considerations, and more.
    - Forward Planning: determine the time needed to complete each action step and when the expected completion date should be.

  - Evaluate key decision points and determine when a move to a contingency plan should be taken.
- Who is the responsible action owner? If the appropriate owners are not included, there is a high risk that the applicable definition of the risk and its evaluation may not be correct.
- What resources are required? For sure cost, and employees to carry out the task of the risk. In addition, a cooperative environment should be encouraged so that all intra departments may contribute as needed, without the fear of intimidation at any level.
- How will this action reduce the probability or severity of impact? This is profoundly important as it depends on the specific definition of the problem. The more succinct the definition is, the more assured the analyst studying the risk will be to identify both severity and occurrence.
- Determining mitigation plans: A mitigating plan is a plan that adapts quickly when risk levels become unacceptable. This becomes a model or a repeatable process that can be used to protect one's business by identifying areas of risk in your business. Issues that are taken into consideration for any kind of mitigation are typically the following:
  - Understand the users and their needs. The users/operational decision makers will be the decision authority for accepting and avoiding risks. Maintain a close relationship with the user community throughout the system-engineering life cycle. Realize that mission accomplishment is paramount to the user community and acceptance of residual risk should be firmly rooted in a mission decision.
  - Seek out the experts and use them: Seek out the experts within and outside your organization for technical experts (as needed) and always provide support in their endeavor to a specific solution. Encourage the team to think "outside the box" for real innovations and drastic changes of continual improvement. Encourage the team to take the time and understand what is feasible and what is not; what is within the resources of what the organization can offer and what is not; what has been implemented either in your organization or in other organizations (if that is the case perhaps the team needs to conduct a benchmarking study). In all cases, management must select appropriate individuals with appropriate knowledge who understand and are able to differentiate between doable and not doable or hard and not so hard work applicable to the risk at hand. Above all, the management must encourage new relations and or cultivate existing relations.

- Recognize risks that recur. This is an inherent issue of a poor problem-solving methodology. Therefore, management should review the methodology used and make sure that they review areas (a) where the risks repeat, (b) interfaces (this is the most common area where things are overlooked), (c) hard points of design or process (this is where even the experts take for granted the notion that everything is OK), (d) third party interests (contractors and consultants), (e) newer environmental and safety requirements that may have a direct or indirect effect on risk(s), and (f) gaps found in the process of auditing or mitigating risk(s) for the particular project. Once it is determined what the issues are, then it is imperative for management to help create an acceptable action plan and monitor them for effectiveness in all areas for the owners.
- Encourage risk taking. There is an old saying that guarantees perfection and no risk of any kind. Fire everyone and shut your organization down, then you will not worry about the headaches that a business faces daily. Obviously, that is absurd! Since we do have employees and an ongoing organization, it is inevitable that problems and risks will arise, and it is up to the organization to face and solve them. Facing problems and risks brings its own consequences, and some of them may be negative. So, what is one to do? The answer is to imitate a turtle—that is, take a calculated risk. Just like a turtle takes a chance and gets out of its shell (risk-taking activity), so we must also take risks—in fact, we must be encouraged to take risks. In the case of the turtle, if she stays in the shell she is guaranteed safety, but she will die of starvation. On the other hand, every time she sticks her neck out of the shell, she is taking a chance that one of many predators will devour her. So, the dilemma for the turtle is a serious one—but in the end, she always takes the risk of not having a predator waiting for her. For the turtle, the risk taking is worth it, and in fact some of them live to be over 150 years old. For a business organization, it is imperative that risk taking should be encouraged by all leadership positions for all high cost products and processes, as well as critical, safety, and environmental issues. After all, all risks are opportunities and they should not be looked upon as discouragement.
- Recognize opportunities. As we just mentioned in the previous point, risk identification and evaluation must be proactive. That means that management should be looking for opportunities for continual improvement in overall performance, capacity, flexibility, or desirable attributes in other areas not directly associated with a given risk.
- Encourage deliberate consideration of mitigation options. This is an excellent approach and consideration for launching any

new product or process. This is so because—especially during launch—things change very fast and an answer is usually competing with other priorities due to tight scheduling. Therefore, it is imperative that a systematic, careful overview of mitigation options and encouragement of thorough discussion by the program team takes place. This is a classic opportunity to "think outside the box."

- Not all risks require mitigation plans. As we all know, not everything is important or critical. Therefore, when looking at risks, we should prioritize them and then look for ways to eliminate, minimize, or find mitigating events to address them appropriately.

The prioritization process should start with a plan for mitigation. Having said this, the priority obviously cannot be initiated without the proper identification of a (a) risk manager, (b) mitigation strategy, (c) implementation and evaluation strategy, and (d) contingency plan.

It is profoundly important to have the risk manager identify and implement any risk mitigation as it applies to a specific risk. To do that effectively, the manager must have appropriate and applicable knowledge, authority, and power to authorize and allocate resources and be in charge of any implementation. Sometimes, depending on the criticality of the risk, it may be necessary to include or escalate issues to higher management level.

The mitigation strategy is a planning task that allows the manager to study and take into account the impact of the severity, as well as the occurrence, of the risk under consideration.

The evaluation and implementation strategy is based on both formative and summative approaches to assessment. The formative evaluation will consider issues that came up during the implementation process, and the summative assessment is the review at the end of the project. Typical issues are: what did or did not work? What we could have done better?

Without a doubt, in dealing with risks of any kind, nobody is sure of what may happen. Therefore, the manager should have created contingency plans to deal with surprises or the "what if" situations that may arise in the process of dealing with risk. Typical actions may be to (a) look for cues that may activate a contingency plan, and (b) review the milestones for the completion of the identified stages in the risk implementation strategy. If the review does not take place, there is a high probability that "at the milestone point" something may not be completed as planned. Vigilance is important here. Therefore, each action must be monitored and checked against the due date of completion and its effectiveness in the total program.

Dealing with risks, quite often people have trouble understanding what exactly the significance of monitoring is for a risk once it has been defined. In its simplest definition, monitoring risk is watching the movement of good, bad or steady behavior of the actions being taken for a given risk. Depending how the action implemented for that risk behaves on or influences the risk, actions will be modified, changed, or discarded all together.

Monitoring risks is a continual (routine review) in management meetings. The foremost questions of the review should be: Are we moving towards progress? Are we improving? Is our target going to be met? Make sure you look at the individual risks identified, rather than focusing on the entire project. Quite often missed deadlines fall apart due to minor incidents that everyone overlooked. To make sure that does not happen, use a program evaluation review technique (PERT) diagram (critical path identification) or a Gant chart schedule.

Part of the responsibility in monitoring risks is to make sure that risks are exposed in a given program (in planning, and during the mitigation process). The idea is *not* to overlook any conditions in the current environment for new risks or modification of existing risks.

Implementation: Implementation follows all of the planned methods for mitigating the effect of the risks. Purchase insurance policies for the risks when it has been decided to transfer the risk to an insurer, avoid all risks that can be avoided without sacrificing the entity's goals, reduce others, and retain the rest.

Risk management plan

Select appropriate controls or countermeasures to measure each risk. Risk mitigation needs to be approved by the appropriate level of management. For instance, a risk concerning the image of the organization should have a top management decision behind it, whereas IT management should have the authority to decide on computer virus risks.

The risk management plan should propose applicable and effective security controls for managing the risks. For example, an observed high risk of computer viruses could be mitigated by acquiring and implementing antivirus software. A good risk management plan should contain a schedule for control implementation and responsible persons for those actions.

According to ISO/IEC 27001, the stage immediately after completion of the risk assessment phase consists of preparing a risk treatment plan, which should document the decisions about how each of the identified risks should be handled. Mitigation of risks often means selection of security controls, which should be documented in a statement of applicability that identifies which particular control objectives and controls from the standard have been selected and why.

Risk options

Risk mitigation measures are usually formulated according to one or more of the following major risk options. Later research (Crockford, 1986, p. 18) has shown that the financial benefits of risk management are less dependent on the formula used but are more dependent on the frequency and how risk assessment is performed. In business it is imperative to be able to present the findings of risk assessments in financial, market, or schedule terms. Courtney (IBM, 1970) proposed a formula for presenting risks in financial terms. The Courtney formula was accepted as the official risk analysis method for US governmental agencies. The formula proposes calculation of ALE (annualized loss expectancy) and compares the expected loss value to the security control implementation costs (cost-benefit analysis).

Potential risk treatments

Once risks have been identified and assessed, all techniques to manage the risk fall into one or more of these four major categories (Stamatis, 2014; Dorfman, 2007):

1. *Avoidance (eliminate, withdraw from or not become involved)*: This includes not performing an activity that could carry risk. An example would be not buying a machine with high life cycle costs, even though the original price of the machine may be less than a machine with low life cycle costs the original price of which may be higher. Avoidance may seem the answer to all risks, but avoiding risks also means losing out on the potential gain that accepting (retaining) the risk may have allowed. Not entering a business to avoid the risk of loss also avoids the possibility of earning profits (McGivern and Fischer, 2012, pp. 289–296).
2. *Reduction (optimize—mitigate)*: Risk reduction or optimization involves reducing the severity of the loss or the likelihood of the loss from occurring. A classic example of this is whether or not to buy water sprinklers designed to put out a fire and therefore, reduce the risk of loss by fire. However, by installing the sprinklers you may cause another loss due to water damage, which may be a higher cost and not suitable for the budget of the organization. The solution may be in finding another alternative. Acknowledging that risks can be positive or negative, optimizing risks means finding a balance between negative risk and the benefit of the operation or activity and

between risk reduction and effort applied. The idea of reduction is for an organization to optimize a particular risk to a level of residual (minimum) risk that is tolerable (IADC HSE).

Yet another approach to reducing risk is by outsourcing it (Stamatis, 2016, 2014; Roehrig, 2006). A typical method is to insure the loss through a third party. By outsourcing the risk, the organization can focus on more important business development issues rather than worry about manufacturing process or some other task within the organization.

3. *Sharing (transfer—outsource or insure)*: Briefly defined as "sharing with another party the burden of loss or the benefit of gain, from a risk, and the measures to reduce a risk."

   The term of "risk transfer" is often used in place of risk sharing in the mistaken belief that you can transfer a risk to a third party through insurance or outsourcing. In practice, if the insurance company or contractor go bankrupt or end up in court, the original risk is likely to still revert to the first party. As such in the terminology of practitioners and scholars alike, the purchase of an insurance contract is often described as a "transfer of risk." However, technically speaking, the buyer of the contract generally retains legal responsibility for the losses "transferred," meaning that insurance may be described more accurately as a post-event compensatory mechanism. For example, a personal injuries insurance policy does not transfer the risk of a car accident to the insurance company. The risk still lies with the policyholder, namely the person who has been in the accident. The insurance policy simply provides that, if an accident (the event) occurs involving the policy holder, then some compensation may be payable to the policy holder that is commensurate with the suffering/damage.

   Some ways of managing risk fall into multiple categories. Risk retention pools are technically retaining the risk for the group, but spreading it over the whole group involves transfer among individual members of the group. This is different from traditional insurance, in that no premium is exchanged between members of the group up front, but instead losses are assessed to all members of the group.

4. *Retention (accept and budget)*: As unthinkable as it may be, the fact is that risk retention is a reality in everything we do. It involves accepting the loss, or benefit of gain, from a risk when it occurs. In our common language we call it *self-insurance*. In other words, we will take a chance that the failure will not happen or if it does, it will not be detrimental to our organization. Risk retention is a viable strategy for small risks where the cost of insuring against the risk would be greater over time than the total losses sustained. All risks that

are not avoided or transferred are retained by default. This includes risks that are so large or catastrophic that either they cannot be insured against or the premiums would be beyond the reach of the organization. Any amount of potential loss (risk) over the amount insured is retained risk. This may also be acceptable if the chance of a very large loss is small, or if the cost to insure for greater coverage amounts is so great that it would hinder the goals of the organization too much.

Ideal use of these *risk control strategies* may not be possible. Some of them may involve trade-offs that are not acceptable to the organization or person making the risk management decisions. Another source of control, may be the approach that the US Department of Defense takes in using the avoid, control, accept, or transfer (ACAT) model.

Assessment

Once risks have been identified, they must then be assessed as to their potential severity of impact (generally a negative impact, such as damage or loss—severity) and to the probability of occurrence. These quantities can be either simple to measure (e.g., in the case of the value of a lost building) or impossible to know for sure (e.g., in the case of an unlikely event, the probability of occurrence of which is unknown). Therefore, in the assessment process it is critical to make the best-educated decisions in order to properly prioritize the implementation of the risk management plan.

Even a short-term positive improvement can have long-term negative impacts. Take the issue of capacity. As capacity becomes limited, new machinery or a new process will be installed. However, in the short term, that capacity shortage may be eliminated, but in the long term it may present other problems such as inventory, new technology, and expensive maintenance and so on. There are many other engineering examples where expanded capacity (to do any function) is soon filled by increased demand. Since expansion comes at a cost, the resulting growth could become unsustainable without forecasting and management involvement in the planning stage.

The fundamental difficulty in risk assessment is determining the rate of occurrence, because statistical information is not available on all kinds of past incidents and is particularly scanty in the case of catastrophic events, simply because of their infrequency. Furthermore, evaluating the severity of the consequences (impact) is often quite difficult for intangible assets. Asset valuation is another question that needs to be addressed. Thus, best educated opinions and available statistics are the primary sources of information. (Sometimes, if mechanical items are of concern, finite element analysis or simulation studies may help.) Nevertheless, risk assessment should produce such information for senior executives of the organization that the primary risks are easy to understand and that the risk management decisions may be prioritized within overall company goals. There have been several theories and attempts to quantify risks. Many different risk formulae exist, but perhaps the most widely accepted formula for risk quantification is *criticality analysis*, which is: rate (or probability) of occurrence multiplied by the severity (impact). The product of the criticality equals the risk magnitude.

Review and evaluation of the plan

Initial risk management plans will never be perfect. This is because we are not capable of foreseeing all risks and all potential problems that may occur in the short or long term. Things happen unexpectedly and when they do, hopefully we learn, adjust, modify, and/or completely change the task, system, or process in the organization. Obviously, practice, experience, and actual loss results will necessitate changes in the plan and contribute information to allow possible different decisions to be made in dealing with the risks being faced.

Risk analysis results and management plans should be updated periodically. There are two primary reasons for this:

1. To evaluate whether the previously selected security controls are still applicable and effective; and
2. To evaluate the possible risk level changes in the business environment. For example, information risks are a good example of a rapidly changing business environment.

Limitations

Prioritizing the *risk management processes* too highly could keep an organization from ever completing a project or even getting started. This is especially true if other work is suspended until the risk management process is considered complete. It is also important to keep in mind the distinction between risk and uncertainty. Risk can be measured by impacts × probability.

If risks are improperly assessed and prioritized, time can be wasted in dealing with risk of losses that are not likely to occur. Spending too much time assessing and managing unlikely risks can divert resources that could be used more profitably. Unlikely events do occur but if the risk is unlikely enough to occur it may be better to simply retain the risk and deal with the result if the loss does in fact occur. Qualitative risk assessment is subjective and lacks consistency. The primary justification for a formal risk assessment process is legal and bureaucratic.

Conclusion

Risk-based thinking is an element in the process approach (AR). That is:

- Risk-based thinking is an input to management review.
- Risk-based thinking is an element in the continual improvement process that is focused on prevention.
- Risk-based thinking has been demonstrated during audits; a risk register is documented with information that validates an organization has done risk-based thinking.

chapter ten

Warranty analysis

When one discusses warranties, one must be cognizant of two terms; although sometimes the terms are used interchangeably, they are quite different and present different challenges on their own. The two terms used are *guarantee* and *warranty*.

A *guarantee* is a formal assurance that certain conditions will be fulfilled, especially that a product will be of a specified quality. On the other hand, a *warranty* is a written guarantee promising to repair or replace an article if necessary within a specified period. In other words, a warranty is a type of guarantee; a product guarantee and a product warranty are basically the same thing—the company undertakes to repair or replace your goods if they go wrong.

The difference is legal, not linguistic. Both terms are meaningful in the context of a contract or bargain. A *guarantee* is a promise that if a thing is not of a certain standard or does not fulfill some condition, the original price or consideration paid for the contract or bargain will be returned. For example, if A sells an item to B and guarantees that it will produce 20 widgets a day, B can return the item to A for a full refund if it does not produce 20 widgets a day. Similarly, C can guarantee A's debt to B, so that if A fails to pay B, C, the guarantor, is obliged to pay B instead.

A *warranty*, on the other hand, is the term for a contract breach that gives rise to a claim for damages, but not the repudiation of the whole contract. For example, if A sells an item to B and warrants that it can produce 20 widgets per day, but in fact it produces only 19, B can bring an action for damages against A for the lesser of (1) the cost of fixing the item such that it does in fact produce 20 widgets per day; or (2) the loss of profit associated with the production of 19 as opposed to 20 widgets. B cannot, however, return the item for a full refund.

In the consumer goods context, where statute provides in many countries for a manufacturer's warranty, a warranty usually connotes that the warrantor may repair or replace a product that has developed a fault during the warranty period as a result of a defect in design or manufacture at the warrantor's discretion. The fact that the warrantor may choose to repair the product and not give a refund is what distinguishes this from a guarantee. The phrase "money back guarantee" is common in the consumer context, and means just that, although the phrase is arguably redundant.

In the manufacturing world, we deal with warranties in a slightly different manner. For example, if company A sells ten widgets to company B and B receives the ten widgets, but for some reason the delivered widgets do not perform to specifications, then B will charge back the appropriate amount for the defective widgets involved. The amount of charge back may be 100% of the cost or a percentage amount that has been agreed upfront.

In a market-directed economy, manufactured products are distributed under contracts that both define the continuing relationship among producers, distributors, and consumers and provide the method by which control of the goods is transferred. The transactions through which goods are transferred by sale are governed, in the main, by Article 2 of the Uniform Commercial Code.

Four theories on the economic rationale for warranty provision have been proposed in the literature: (1) warranties provide insurance to customers and work as a risk-sharing mechanism, (2) warranties are a sorting mechanism and work as a means for second-degree price discrimination among customers with different risk preferences, (3) warranties work as a signal of product quality to consumers under information asymmetry, and (4) warranties work as an incentive mechanism for firms to reveal and improve product quality.

In the modern world we have come to expect certain functions of the "product" or "service" to last a definite period of time. When that does not happen, a warranty covers the prorated usage and makes good the original contract with the customer. That process is called liability to the manufacturer or service provider. There are four theories of liability commonly used in defective product cases. Understanding these theories will help you decide if you want to pursue a defective product claim. Keep in mind that you can use one or more of these theories at the same time. Here's a brief overview of the theories.

Breach of express warranty: An "express warranty" is any type of warranty or guarantee that is written or stated. Such written statements can be:

- On the product's label or packaging
- In the instructions or other paperwork included with the product
- On signs or other marketing materials at the store where you purchased the product
- In any form of advertising for that product

Any of these representations about the defective product may be an express warranty.

Breach of implied warranty: If the defective product you used did not come with an express warranty—and even if it did—that product may be covered by implied warranties, and the defect may have violated those implied warranties. An "implied warranty" is a warranty that the law automatically applies to your product—it doesn't have to be guaranteed by the manufacturer or store where you bought the defective product. State law imposes these warranties on product manufacturers and suppliers, whether they like it or not.

The implied warranties that apply to your case will depend on the particular product involved and the circumstances surrounding its sale. Though they may differ somewhat from state to state, implied warranties generally come in two forms:

1. **Implied warranty of merchantability.** This is a guarantee that a product is reasonably fit for the purpose for which it was sold.
2. **Implied warranty of fitness for a particular purpose.** This imposes an additional obligation in cases where a seller knows that the buyer of a product intends to use it for a particular purpose. This warranty is an additional guarantee that the product will be reasonably fit for that purpose.

Strict product liability: If there is an "ace in the hole" in defective product cases, it is the legal doctrine of strict liability. Here's how strict product liability works: normally, if a company is the cause of an accident of some sort, the company will only be held liable if it is found to have acted negligently—in other words, the company did not take normal care or precautions in its actions (called the failure to "exercise reasonable care"). With strict liability, however, the company will be liable regardless of the care it exercised or the precautions it took to prevent an accident.

As a practical matter, strict liability means that when you present your defective product claim, you don't need to show that the manufacturer or supplier of the defective product was not sufficiently careful in making or distributing that product. You just have to show that the product is somehow defective and that the defect was the cause of your injury.

In opposing your defective product claim, the defendants in your case may argue that the product was not unreasonably dangerous, that you should have been aware of the danger and avoided it, or that the defect was not the cause of your injury. If you are able to base your claim on strict product liability, however, the defendants will not be able argue that they were really, really careful when they made or distributed the defective product. Strict product liability is a rapidly developing area of the law, and its application varies considerably from state to state. Most states have adopted some version of strict product liability, but it may not be available in every case.

Negligence: In addition to a claim based on strict product liability, or in cases or jurisdictions in which strict liability is not an available legal basis for your claim, you may be able to argue that defendants acted negligently in manufacturing or supplying the defective product that injured you.

In order to prove negligence, you must show that the defendants were not reasonably careful (called "failing to exercise reasonable care") in making or distributing the injury-causing product. Proving negligence can be difficult—but the difficulty varies depending on the type of product and manufacturing involved.

Intentional misrepresentation or fraud: In some cases, the evidence may show that a defendant knew of a dangerous defect associated with a product and deliberately concealed the danger or marketed the product using deliberately misleading statements. In such cases, you may have a claim for intentional misrepresentation or a tort claim based on the fraudulent conduct (the names may vary from state to state).

If the defendant is a corporation, what the defendant "knew" may be based not on information hidden in someone's brain (or memory), but rather on information contained in the company's records, such as FMEAs, control plans and APQP documentation.

Automotive perception of warranty

Warranty maybe thought of as customer dissatisfaction. It is a multi-billion dollar a year opportunity for organizations: it is an opportunity to know about warranty, understand it, and apply its principles so that the organization can have *less warranty, which will result in improved profits.*

Obviously, not everything may have a warranty. For example, items that are generally excluded are:

- Normal wear and tear
- Maintenance (oil and fluid changes, tune-ups, wiper blades, oil or air filters, clutch linings, etc.)
- Damage from environment or weather
- Inappropriate or poor maintenance (incorrect fuel, fluids, etc.)
- Damage from accidents

Within the automotive industry, one may come across two distinct types of warranty. However, even though they are very similar, it must be noted that they occur at different times. The two types are:

1. *Warranty cost*: This is an estimate of the lifetime warranty obligation on a new vehicle at the time of sale. The cost estimate is accrued, increasing the warranty reserve. This of course affects company profits as reported in the income statement. Another way of thinking about this, is to think of putting money in a savings account to pay for future warranty claims.

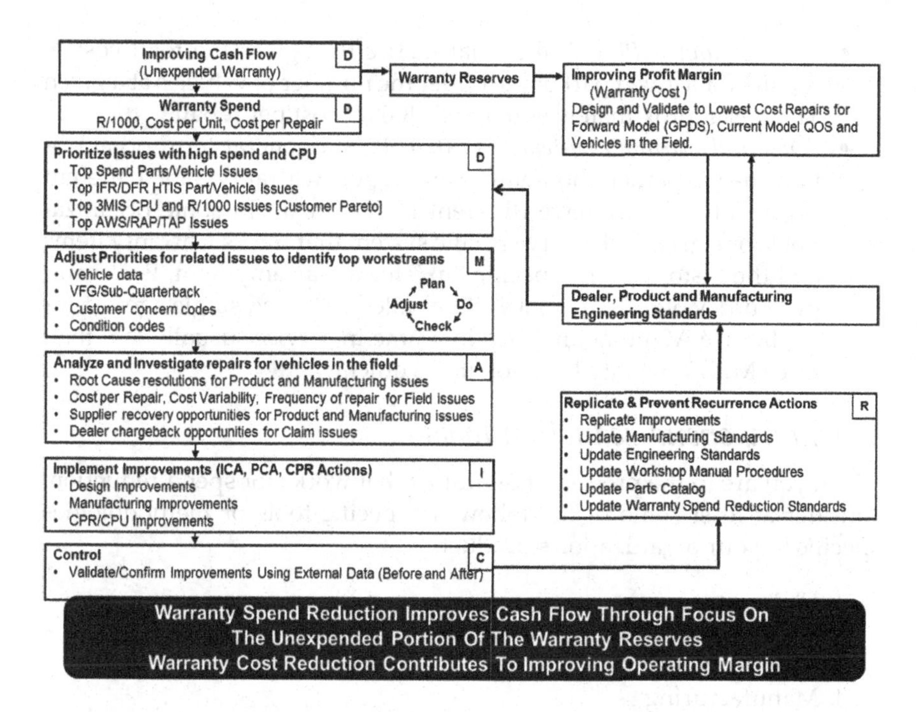

Figure 10.1 Warranty spend improvement (D = Do, M = measure, A = Analyze, I = implement, C = Control, R = replicate).

2. *Warranty spend*: This is the *actual* occurrence of warranty obligations as they are claimed. The warranty spend is paid, decreasing the warranty reserve. This of course affects company cash flow as reported in the balance sheet. Another way of thinking of this, is to think of withdrawals from the savings account when bills are due. A typical spend improvement flow is shown in Figure 10.1; thusly (Jocz, 2018):

Typical terms used in the auto industry:

- *Months in service (MIS) or time in service (TIS)*: Both terms are interchangeable, indicating the number of months that the vehicle has been in the customer's hands; this generally starts with the date of sale of the vehicle to the end customer (it does not include the time sitting on the dealership lot).
- *Repair per 1000 vehicles (R/1000)*: The number of repairs made on a vehicle divided by the number of vehicles made at a given TIS, multiplied by 1000.

- *Cost per unit (CPU)*: A financial metric that gives the total cost of repairs for a given part or parts, per the number of vehicles at a given TIS. It is approximately equal to R/1000 × cost per repair.
- *Analytical warranty system (AWS)*: This is a general term organizations use to gather and analyze and report warranty data. Different organizations may have different names for it. In some organizations, one may find a very specific system that tracks warranty items and the history of the specific "fix" for a warranty item. For example, Ford Motor Co. uses a system called balanced single agenda for quality (BSAQ) that analyzes low time in service (usually less than three MIS) warranty to action items quickly.

Warranty spend reduction tools

When you are in the process of defining what works for spend reduction, consider at least some of the following specific tools or methodologies specific to your organization's product:

1. Claims processing
2. Warranty counseling process
3. Supplier recovery
4. Manufacturing
5. Early detection
6. Special service diagnostics/repair tool
7. Optimize diagnostic/repair procedure
8. Prior approval
9. Field service actions
10. Communications
11. Remanufacturing
12. Service part kit pricing
13. Service part level

For maximum benefit consider developing a formal systemic process(es) for recovery, such as:

- Warranty reduction program (WRP)
- Parts return analysis & supplier chargeback
- Spike recovery process
- Field service action (FSA)
- Warranty special contracts

Perhaps the best *warranty spend reduction* is a methodology for early detection. This methodology may be specific to the organization or product or it may be generic to include potential opportunities to increase tool use, optimize tool use, and create additional tools to *ensure identification of concerns as early as possible*. Some methodologies or tools to consider are:

- Overall awareness of how warranty lags other metrics.
- Ensure that warranty is taken into account while making or sourcing evaluations/decisions.
- Use a supplier warranty reporting (SWR) tool to familiarize yourself with all pertinent warranty performance.
- Review warranty improvement plans when on-site with appropriate personnel.
- Drive closure of open warranty issues and ensure that there is linkage to the supplier's quality operating system (QOS).
- Seek input from the customer's representative on your performance and ensure all issues are addressed systemically via the manufacturing site assessment (MSA).
- Ask your customers or suppliers to provide examples of their warranty performance and actions they have taken to improve performance.

Sometimes it is necessary to have the customer representative intervene directly with the supplier so that he or she can influence the supplier's QOS and make significant contributions to warranty improvements either on current or future products by identifying key matrices and suggesting appropriate monitoring. Intervention may encompass a variety of issues and circumstances, such as:

- Engage on supplier process (SP) and support supplier design (SD) in a formal and standardized process or methodology to ensure timely closure.
- Support any weak current or potential issues or review any pending plant actions with respect to suppliers.
- Drive supplier improvements using the MSA warranty validation areas, including discussion of how supplier is managing warranty for their parts.
- Review and evaluate warranty spike (unusual or unique events) recovery to find assignable causes for the increase in warranty and propose reasonable solutions.
- Help the supplier develop a standardized organizational warranty chargeback system and warranty reduction program—see Table 10.1.

Table 10.1 Typical warranty programs

Program	Warranty reduction program (WRP)	Global warranty chargeback (GWC)	Warranty spike recovery (WSR)
What	Incentive-based program Tool: Management review; benchmarking;	Ongoing part reimbursement for supplier part failures Tool: Cost of quality; financial analysis	Special cause recovery Tool: 8D; CAPA methodology; audit
Why	A *No Fault* approach to look at total spending and liability Tool: Cost of quality;	Financial accountability for supplier responsible failures Tool: QOS; cost of quality; management review	Financial accountability for supplier responsible failures Tool: QOS; cost of quality; management review
How	MY to MY variance in absolute warranty based on WRP. (MY = Month, Year) Tool: ANOVA; t-test	Responsibility assessed through analysis of failed warranty parts Tool: Weibull analysis; tests to failure	Responsibility assessed through various means Tool: Audit; cost of quality; management review

A typical starting point for developing a standardized warranty model is to start with a specific commodity and include at least three items:

1. Current warranty performance
2. Commodity characteristic questions (e.g., difficulty of determining root cause; supplier design responsibility and system interaction issues; use a P-diagram, interface diagram, FMEAs; if mechanical item, use finite element analysis, etc.)
3. The ability to determine warranty performance targets sourcing and supplier cooperation level. The intent of this item is to develop a practical (doable) warranty reduction program. Tangible results could be:
   a. Customer warranty and product development quality will assist the selection of suitable components/suppliers and the required WRP agreements. They may even help with appropriate GD&T usage.
   b. WRP agreements can be entered into at any point in time.

c. Warranty targets are either pre-set for a MY (fixed, for agreements in line with a specific date) or based on previous MY performance (floating, for existing agreements in line with previous WRP guideline).
d. Warranty performance versus target can be established at various stages during the coverage period (also called pay-point).

chapter eleven

GD&T used in the APQP and PPP process

The focus of GD&T is to provide the customer with what they want, when they want the product, how or the form they want the product in and at the price they want (Stamatis, 2016, pp. 305–334; ASME Y14.5, 2009; ISO 16792, 2015).

What is geometric dimensioning and tolerancing (GD&T; also GDT)?

- GD&T is a dimensional control of solid products.
- Above all, it is a uniform system that can be understood worldwide.
- GD&T is the universal language communicating the engineering wants in a visual form to all key players from the concept of customer-required controls to gauging (verification) of controls.
- It is a mechanism providing dimensional certainty that leads to proper fit and function.

How does GD&T work?

- GD&T establishes central controls that cover the entire dimensional part—it is a visual language.
  - It establishes foundational features, called "datums," which are based on part features that typically mate to other parts.
  - It establishes characteristic controls that provide tolerance to part features through an instruction called the feature control frame (FCF).
  - Dimensions define features of size (FoS) and locate features to the datum structure.
  - The final result is a dimensional control system that allows parts to be built into assemblies, allowing variation with respect to the "best" dimensional stability each part offers to the system; i.e., it tries to avoid or minimize issues with "stack-up tolerances."

Geometric characteristic symbols: GD&T is an international visual language. As such, the symbols used are defined by the ASME Y14.5-2009 standard. They are grouped in the following matrix based on the nature of their control and whether they are related to their features. Table 11.1 shows some of the most common symbols used in GD&T. For the entire list of symbols, see ASME Y14.5-2009 Section 8.3.1.2.

Table 11.1 GD&T essential definitions, based on ASME Y14.5-200 Section 8.3.1.2 and ISO 16792 3D

For individual features	Form	Straightness
		Flatness
		Circularity
		Cylindricity
For individual or related features	Profile	Profile of line
		Profile of surface
For related features	Orientation	Parallelism
		Perpendicularity
		Angularity
	Location	Position
		Concentricity
		Symmetry
	Runout	Circular Runout
		Total Runout

GD&T essential definitions

Form

Straightness: a condition where an element of a surface or an axis is a straight line.

Flatness: the condition of a surface having all elements in one plane.

Circularity: (1) when applied to cylinder or cone, a condition where all points on the surface intersected by any plane perpendicular to the axis are equidistant from the axis; (2) when applied to a sphere, a condition where all points of the surface intersected by any plane passing through a common center are equidistant from that center; (3) circularity zone, two concentric circles, radial separation equal to the circularity tolerance value, within which the considered circular element must lie.

Cylindricity: a condition of a surface of revolution in which all points of a surface are equidistant from a common axis.

Profile

Profile of a line: the condition permitting a uniform amount of profile variation, either unilaterally or bilaterally, along a line element of a feature.

Profile of a surface: the condition permitting a uniform amount of profile variation, either unilaterally or bilaterally, on a surface.

Orientation

Parallelism: the condition of a surface, line, or axis that is equidistant at all points from a datum plane or axis.

Perpendicularity: the condition of a surface, axis, or line that is 90° from a datum plane or axis.

Angularity: the condition of a surface, axis, or center-plane that is at a specified angle from a datum plane or axis.

Location

Position: a zone within which the axis or center plane of a feature is permitted to vary from the true (theoretically exact) position.

Concentricity: a condition in which two or more features, in any combination, have a common axis.

Symmetry: a condition in which a feature (or features) is (or are) symmetrically disposed about the center plane of a datum feature.

Runout

Circular runout: the composite deviation from the desired form of a part surface of revolution through a full rotation (360°) of the part on a datum axis.

Total runout: the simultaneous composite control of all elements of a surface at all circular and profile measuring positions as the part is rotated through 360°.

FoS: either a cylindrical or a spherical surface or a set of two opposite parallel surfaces with a dimension. It must meet two requirements: (1) it should be a spherical, cylindrical, or a parallel set of surfaces for which a center axis or plane can be calculated from the surfaces; and (2) the feature should be associated with a dimension.

RFS: regardless of feature size (RFS) is a concept where no material modifier symbol is shown. That means that the tolerance applied to a feature is through its entire range of size from MMC to LMC.

MMC: maximum material condition is the condition where the part weighs the most. It also means that the external features, like a dowel pin or a bolt, will be at their largest permissible diameters, or it may be that the internal feature, like a hole, will be at its smallest permissible diameter, i.e., MMC makes the part the heaviest.

LMC: implies that condition of a part feature of size wherein it contains the least (minimum) amount of material, e.g., largest hole size and smallest shaft size. It is opposite to MMC: i.e., LMC makes the part the lightest.

Modifier: is helpful for clearance fits. It allows the tolerance to increase as the size of the feature varies. It can also be used on datum references if there might be looseness or "play" on those datum features.

Datum: A datum is a virtual ideal plane, line, point, or axis. A *datum feature* is a physical feature of a part identified by a *datum feature symbol* and corresponding *datum feature triangle*, e.g., A—◀

FCF: The FCF provides an orderly presentation of the geometric control, tolerance, tolerance zone shape, tolerance modifiers, and any datum reference(s) if they are required. A typical FCF is the following:

A	B C	D E	F	G

A is reserved for the geometric characteristic symbol. B is reserved for tolerance; if it is a cylindrical tolerance, a diameter symbol precedes the tolerance, and if it is a width, no symbol is used. C is where a modifier symbol may be added next to the tolerance to indicate the tolerance applies at MMC or LMC. D is where datums may be added if this FCF is controlling something other than one of the four form characteristics and other exceptions. E is where, when required, the datum can be shown as applying at a stated boundary condition of MMB or MLB. F is where additional datum points may be added in the FCF to reference secondary or tertiary datum (G). *Important note*: The alphabetical or numerical order *does not* define the datum priority; the order is always defined by the letter or number in the box location.

GD&T rules

- Rule #1: Envelop principle
  - When a regular FoS is at MMC, it cannot extend beyond the boundary (envelop) of the perfect form at MMC.
- Exceptions: the boundary of the perfect form at MMC does not apply to
  - Stock sizes—materials produced to industry or government standards,
  - Parts with a straightness or flatness tolerance added to the size dimensions,
  - Part that invoke the independency symbol (I), and
  - Non-rigid parts, such as O-rings, molded or body panels (these may be controlled with some form of restricted inspection.

Note: If geometric controls are not specified, the drawing is incomplete and parts may be produced with unspecified (unintended) results. After all, perfect form at MMC does not mean perfect orientation or location. Orientation and or location may be specified and assured with poka-yoke—error (design) or mistake (process) methods.

- Rule #2: Applications of modifiers on geometric tolerance values and datum features references
  - RFS on features on features of size, if MMC or LMC is not recognized;
  - Regardless of material boundary (RMB) on datum references that are a FoS, where maximum material boundary (MMB) or least material boundary (LMB) is not specified;
- Fit and function
  - Fit the condition where solid parts, when assembled, are not predisposed to occupy the same space at the same time;
  - There are various functions in the assembling of parts such as the clearance of parts, joint integrity, part wall thickness protection, alignment, and balance;
  - Simple functions involve dimensional controls that may result in simple manufacturing methods; and

- More complex functions involve more sophisticated controls and may result in more complex manufacturing methods.

Things to remember about GD&T

1. Form
   a. No datum reference allowed.
   b. Only one feature may be controlled with Form.
   c. A seal groove may create two surfaces and profile will need to be used instead of flatness.
   d. Straightness tolerance of an axis RFS and at MMC can override Rule 1.
   e. Of all the form controls, only straightness tolerance of an axis can use the MMC modifier.
   f. However, only attribute data will be obtained from the gauge (fits or doesn't fit).
   g. Don't use form tolerances greater than the limits of size when specifying flatness at RFS.
2. Orientation
   a. Orientation controls do not provide location for features.
   b. Location and profile controls or tolerance dimensions are the only ways to locate surfaces and features.
   c. Orientation tolerances should be significantly smaller than the position tolerance, in order to make it a cost-effective control.
   d. Make sure sufficient datums are referenced for full orientation control.
3. Position
   a. Position is the measurement of permissible variation from theoretical design intent (exact location).
   b. Location of one or multiple FoS is relative to one another or to specified datums.
   c. Basic dimensions establish the theoretical design intent (exact location).
   d. Positional tolerance applies only to the FoS location, *not* to the basic dimensions.
   e. Positional tolerancing is applied at RFS, MMC, and LMC.
   f. Axis and surface interpretations:
      i. Axis: zone defining variation allowance of center, axis or center plane of a FoS from an exact location.
      ii. Surface: a VC boundary, located at an exact location that may not violate the FoS boundary.
4. Profile
   a. Profile can be used with or without referencing datums.
   b. The profile tolerance is applied RFS. Currently, no modifiers are allowed in the tolerance zone with the exception of Circle U.

c. Unless otherwise specified, the profile tolerance is equal bilaterally.
d. The profile tolerance can also be unilateral or unequally disposed by adding support geometry showing the amount of tolerance that this applied outside of the part or by using the Circle U modifier.
e. Referenced datums are allowed to have material modifiers, yet the added tolerance needs to be considered.
f. Remember—"Shift Happens!" Sometimes due to tool ware sometimes unexpectedly!
g. To minimize the effects of datum shift, particularly when the datum features have a large size tolerance, reference datums at RMB or decrease the range of the datum feature size tolerance.

- Datums
  a. Always verify that the selected datum is a real feature that can be measured.
  b. Verify that the selected features have form and size controls applied.
  c. Ensure that the primary, secondary, and tertiary datum features been related with location and orientation controls.
  d. Verify that the features selected are permanent. If the features are not permanent, determine how the relationship will be measured when they are gone.
  e. Datum targets may require additional datum references to ensure repeatability.

Datums are required for gauge studies and layouts. They help in conducting:

- Repeatability
- Reproducibility
- Stability
- Datums are also required to establish foundational dimensional controls for a part or assembly

Core team

With the proliferation of globalization, it is not unusual to have virtual teams working on design issues. Virtual teams in today's organizations consist of employees working at home and small groups in the office but in different geographic locations. GD&T in many organizations falls in this category. In the strictest sense, a core team is made up of about five to nine individuals who have appropriate and applicable knowledge, authority, and responsibility to carry out the decision(s) of the team and, above all, must have ownership of what is being discussed. As the work progresses, other individuals may enter the discussion, based on necessity or attrition of members. Some of the considerations for "good" teams are the following.

- Select the best employees, who may be located anywhere in the world.
- Recognize that workers demand personal flexibility. When flexibility is offered, workers tend to be more productive—for example, less commuting and travel time. Changes in workers' expectations of organizational participation may also contribute to productivity.
- As technology changes, be aware that workers will demand increasing technological sophistication. A flexible organization is more competitive and responsive to the marketplace.
- The increasing globalization of trade and corporate activity dictates recognition that the global workday is 24 versus 8 hours, as well as the emergence of environments that require inter-organizational cooperation in addition to competition.
- A continued shift from production to service/knowledge work environments, demands a cross-functional and multidiscipline makeup of the team. This is very difficult, but is necessary in our global world. Human resources are increasingly participating in horizontal organization structures characterized by structural and geographical demands to deliver excellence in the global organization.

The structure of a basic virtual team may be in one of the following seven types:

1. *Networked Teams* consist of individuals who collaborate to achieve a common goal or purpose; membership is frequently diffuse and fluid.
2. *Parallel Teams* work in the short term to develop recommendations for an improvement in a process or system and have a distinct membership.
3. *Project or Product-Development Teams* conduct projects for users or customers for a defined period of time. Tasks are usually non-routine, and the results are specific and measurable. The team has decision-making authority. This is the type of team most used in a GD&T environment.
4. *Work or Production Teams* perform regular and ongoing work, usually in one function and have clearly defined membership.
5. *Service Teams* support customers or the internal organization, typically in a service/technical support role around the clock.
6. *Management Teams* work collaboratively on a daily basis within a functional division of a corporation.
7. *Action Teams* offer immediate responses activated in (typically) emergency situations.

Steps for effective GD&T

Just like most endeavors in engineering and manufacturing, GD&T methodology is a team effort. To be effective in this application, at least five steps must be defined and implemented.

1. *Create a need and transfer that need to a CAD model of the proposed design.* The part is modeled in CAD in its perfect form. Tolerances are not determined at this point.
2. *Assemble a team and decide the best datum structure and relate all features to the datums with basic dimensions.* Decide what features may be used as datums. Remember: parts have multiple features, so engineers have several choices for datums. A simple approach may be to consider what is functional, i.e., is the assembly or manufacturing capable of producing the design as planned? Some additional considerations may be whether this is a new design or are the mating parts carried over from a previous level? How does the part function? What surfaces or features mate to other components? Does any surface seal to other surfaces? What features need to be oriented or positioned to other features on this part? What features must be oriented to the mating part(s)?
   a. The functional mating feature becomes the primary datum. The team determines which features are functional to the part when it is assembled to other parts (components). These datums are the basis for future functional relationships and measurements.
   b. What feature or features locate? This becomes the secondary datum.
   c. What features or features stop rotation? This may define the tertiary datum.
3. *Establish GDT controls.* This is an essential consideration as it will define the primary datum, which, as an example, may need flatness or profile. Is the perfect form at MMC? The secondary datum will require orientation or positioning to the primary datum. Finally, the tertiary datum will require positioning to the primary and secondary datum. Typical controls may be: (a) positional tolerances referencing the datums for holes and (b) the profile of a surface, referenced to the datums, that may be selected to control the outside of the part.
4. *Assign tolerances based on component interfaces.* A very fundamental question for a primary datum feature is to answer the question: what flatness or profile is required for this mating surface? The fundamental question for a secondary datum feature is to answer the question: have the mating parts been examined as to whether or not the perpendicularity tolerance has been determined? At this point of GD&T evaluation, the "stack up" analysis should be completed and the appropriate parameters should be in place.

5. *Reassemble the team for approval.* The idea here is to review the design for applicability and appropriateness. As such:
   a. Discuss the completed part with the entire team.
   b. Develop a gauging strategy (if one has not been defined). If it has been defined, review that strategy and the control plan that is associated with it.
   c. Gauge drawings may also be reviewed here.
   d. Critical areas may be identified and captured by the control plan, touch point models, or CAD data.

chapter twelve

Managing change

Overview

In the APQP process there is an inherent change in every area of the process. Therefore, the management leadership must be aware of at least four items: (a) what the changes are, (b) what has to be done, (c) the execution strategy for satisfactory completion,, and (d) the maintenance of that strategy for the long run.

These four items are obviously driven by individuals in leadership (management) positions. It depends on their personal characteristics whether they can be winners, hold onto the status quo, or be losers. The three key characteristics that propel leaders to take action are based on their own style for managing change. Generally, there are three styles.

1. *Conserver*: This type of manager always prefers the status quo. They feel comfortable with the present situation—maybe with very, very small changes. They are uneasy with unscheduled and unscripted situations because they are very disciplined, deliberate, and detailed in their approach.
2. *Pragmatist*: This type of manager is middle of the road, but also likes to explore new situations with low levels of risk(s). This approach allows the manager to be seen as a risk taker, although the risk is limited based on objective, reasonable, practical, agreeable, and flexible approaches. This management style is very conscious of not "rocking the boat" in the organization.
3. *Originator*: This type of manager is always *pushing the envelope* of success. They are relentless in pace and offering options. Quite often they are unconventional, spontaneous, and willing to take risk(s). They seem to challenge the status quo, because they inherently believe something is *always* or *could be* better.

Regardless where a manager fits in the styles just mentioned, it is important to recognize that all of us may not be 100% exclusively bonded to one style. Quite often we flow in our styles from one category to another depending on the organization and the specific issue.

On the other hand, it is also equally important to recognize some fundamental issues, regardless of style. In the big picture, these fundamentals are divided in two categories, identifying strengths and identifying weaknesses, of any organization upon implementation of a specific change. Specific time should be allotted for an open discussion in the organization about the changes needed to satisfy customer requirements. Some of the considerations should be:

1. *Change awareness*: The organization's ability to proactively search for and see opportunities for renewal and innovation (the optimum goal of APQP).
2. *Change agility*: The capacity of the organization's leaders to facilitate and deliver change that is needed (the optimum goal of customer expectations).
3. *Change reaction*: The ability of an organization's people to read and respond to change that is not planned (the optimum goal of the organization and customer).
4. *Change mechanisms*: The organization's structures and systems that support the implementation of change (the optimum goal of the organization).

Managing change

Let us all remember that *if we always do what we always did, we will always get what we always got!* This means that if we are not willing to embrace change, we are not ready to lead. Put simply, leadership is not a static endeavor. The need for change exists in every organization; the question is: are the leaders prepared for it? Are they willing to identify the need for change? If not, one cannot effectively lead change without understanding the landscape of change. There are three typical responses to change:

1. *The victim*: Those who view change as a personal attack on their persona, their role, their job, or their area of responsibility. They view everything at an atomic level based upon how they perceive change will directly and indirectly impact them.
2. *The neutral bystander*: This group is neither for nor against change. They will not directly or vocally oppose change, nor will they proactively get behind change. The neutral bystander will just go with the flow, not wanting to make any waves, and thus hoping to perpetually fly under the radar.
3. *The critic*: The critic opposes any and all change.

To avoid these typical responses, managing change requires leaders and management to have control over four critical elements:

1. *Vision alignment*—those that understand and agree with your vision must be leveraged in the change process. Those who disagree must be converted or have their influence neutralized.
2. *Responsibility*—your change agents must have a sufficient level of responsibility to achieve the necessary results,
3. *Accountability*—your change agents must be accountable for reaching their objectives, and
4. *Authority*—as a change agent you must have the power or right to give orders, make decisions, and enforce consequences for those who do not follow the orders.

Obviously, to apply change of any kind one must have a plan—a model of sorts. There are many models one can follow. However, the following three approaches to change are significant and common in most models.

1. **Change management models** have been developed based on research and experience on how to best manage change within an organization or in your personal life. Most change management models provide a supporting process that can apply to your organization or personal growth.
2. **Change management processes** include a sequence of steps or activities that move a change from inception to delivery. It is the sequence of steps or activities that a change management team or project leader follow to apply change management to a change in order to drive individual transitions and ensure the project meets its intended outcomes (see the descriptive model in Table 12.1 and the visual model in Figure 12.1, which are based on Kotter's (1996) model).

Table 12.1 Kotter's descriptive model of change

Change management process	
The people side of change	Business need
• Awareness • Desire • Knowledge • Ability • Reinforcement	• Concept and design • Implementation • Post-implementation

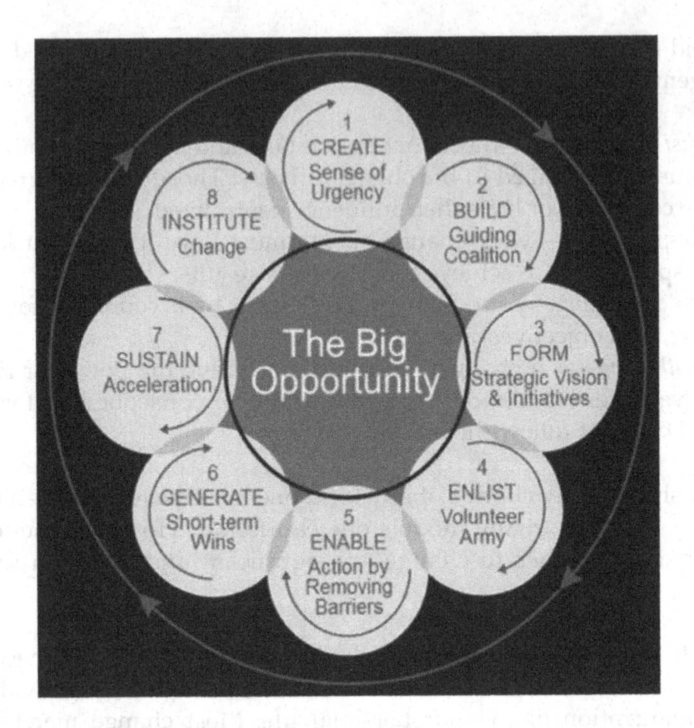

Figure 12.1 Kotter's visual model of change.

3. **Change management plans** are developed to support a project to deliver a change. It is typically created during the planning stage of a change management process.

To ensure the change

1. *Identify what will be improved*: Since most change occurs to improve a process, a product, or an outcome, it is critical to identify the focus and to clarify goals. This also involves identifying the resources and individuals that will facilitate the process and lead the endeavor. Most change systems acknowledge that knowing what to improve creates a solid foundation for clarity, ease, and successful implementation.
2. *Present a solid business case to stakeholders*: There are several layers of stakeholders, including upper management who both direct and finance the endeavor, champions of the process, and those who are directly charged with instituting the new normal. All have different expectations and experiences and there must be a high level of "buy-in" from across the spectrum. The process of onboarding the different constituents varies with each change framework, but all provide plans that call for time, patience, and communication.

3. *Plan for the change*: This is the "roadmap" that identifies the beginning, the route to be taken, and the destination. You will also integrate resources to be leveraged, the scope or objective, and costs into the plan. A critical element of planning is providing a multi-step process rather than sudden, unplanned "sweeping" changes. This involves outlining the project with clear steps that have measurable targets, incentives, measurements, and analysis. Always be prepared to allocate new resources or reallocate current resources. Focus on optimization!
4. *Provide resources and use data for evaluation*: As part of the planning process, resource identification and funding are crucial elements. These can include infrastructure, equipment, and software systems. Also consider the tools needed for re-education, retraining, and rethinking priorities and practices. Many models identify data gathering and analysis as an underutilized element. The clarity of clear reporting on progress allows for better communication, proper and timely distribution of incentives, and measuring successes and milestones. Remember, if you do not have data you cannot measure; if you cannot measure you are guessing. Be data driven for your decisions!
5. *Communication*: This is the "golden thread" that runs through the entire practice of change management. Identifying, planning, onboarding, and executing a good change management plan is dependent on good communication. There are psychological and sociological realities inherent in group cultures. Those already involved have established skill sets, knowledge, and experiences, but they also have pecking orders, territory, and corporate customs that need to be addressed. Providing clear and open lines of communication throughout the process is a critical element in all change modalities. Learn the language of the organization and the people who are working for the change, as well as the people who will implement and be part of it. This demands methods of transparency and two-way communication structures that provide avenues to vent frustrations, applaud what is working, and seamlessly change what doesn't work. Encourage participation and eliminate fear or intimidation for sharing opinions.
6. *Monitor and manage resistance, dependencies, and budgeting risks*: Resistance is a very normal part of change management, but it can threaten the success of a project. Most resistance occurs due to fear of the unknown. It also occurs because there is a fair amount of risk associated with change—the risk of impacting dependencies, return on investment risks, and risks associated with allocating budget to something new. Anticipating and preparing for resistance by arming leadership with tools to manage it will aid in a smooth change

lifecycle. One of the best ways to avoid or minimize risk is to encourage two-way communication.

7. *Celebrate success*: Recognizing milestone achievements is an essential part of any project. When managing a change through its lifecycle, it's important to recognize the success of the teams and individuals involved. This will help in the adoption of both your change management process as well as the adoption of the change itself. It will also be a positive metric for future change plans.
8. *Review, revise and continuously improve*: As much as change is difficult and even painful, it is also an ongoing process. Even change management strategies are commonly adjusted throughout a project. Like communication, this should be woven through all steps to identify and remove roadblocks. Like the need for resources and data, this process is only as good as the commitment to measurement and analysis.

It is often said that *change* is life's only constant. While some changes may be for the better and others may be for the worse, the automotive industry wants everyone to think of change in terms of *risk*. Instead of "change management" think of it as *risk management*. Risk management comes in a variety of forms within a manufacturing operation and the automotive QMS standard adds additional requirements to fully encompass sources for change. Note that the following clauses deal with different aspects of an organization, but all fundamentally revolve around this concept of managing risk when change occurs.

- 16949: 8.5.6.1 Control of changes—supplemental
- 16949: 8.3.6.1 Design and development changes—supplemental
- 16949: 7.5.3.2.2 Engineering specifications

The ability to manage changes within an organization is only as good as the organization's risk management process to identify potential sources of change. The automotive QMS standard, IATF 16949, is very prescriptive in presenting potential sources of change.

In no uncertain terms, IATF 16949 requires the organization to consider the manufacturing facility's support sites within the scope of their QMS. This helps assure a better alignment between the facility's requirements towards the end-customer and the support site's output to the facility. IATF 16949 also requires the organization to consider the customer's requirements within the scope of their QMS. This helps create a complete picture of a QMS that is inclusive of all potential sources of change. Failing to consider any aspect of change introduces risk, and failing to manage and mitigate this risk is failing to meet the requirements of IATF 16949.

To ensure the concept of managing and mitigating risk is carried out throughout the organization, the standard imposes additional awareness requirements. These requirements help ensure that everyone understands not only what they do, but how that interacts with the organization's assessment of risk.

- 16949: 4.3.1 Determining the scope of the quality management system—suppler
- 16949: 4.3.2 Customer-specific requirements
- 16949: 7.3.1 Awareness—supplemental

A final major source of variation and change can come from within the facility itself. Equipment can go down during production and require the use of an alternative station—otherwise known as a "bypass" process. The automotive industry identifies the alternative/bypass process as a potential risk to be developed and addressed during program development. For example, consider a scale that weighs propellant generator; during the course of production, that scale goes down, and the organization instead decides to load by hand using a manual scale. These sorts of changes are taken very seriously by the automotive industry, and the new standard adds a specific clarification to this effect in:

- 16949: 8.5.6.1.1 Temporary change of process controls

System structure

The IATF 16949 identifies a number of areas that the organization should give specific additional consideration. In some cases, these additions augment the basic structure of the QMS and may require additional documented information. In other cases, these additions impact the scope and reach of requirements for control of the supply chain, or close the loop in the identification of potential risk through the analysis of field failures (potential warranty).

In terms of documented information, IATF 16949 requires some specific documents for the QMS, including a quality manual. However, despite being specific about what the quality manual must include at a minimum, the standard does allow flexibility in the format and structure of the document(s) and the way it is stored (electronic or hard copy).

- 16949: 7.5.1.1 Quality management system documentation

Additional system structure is needed to cover automotive-specific items related to supply chain management. These additional items align with the plan–do–check–(study)–act model and deal with the control of suppliers on an operational level. These additions are quite prescriptive and deal with the identification of risks and the level of control to be applied based on the risks identified.

- 16949: 8.4.1.2 Supplier selection process
- 16949: 8.4.2.1 Type and extent of control—supplemental
- 16949: 8.4.2.2 Statutory and regulatory requirements
- 16949: 8.4.2.3.1 Automotive product-related software
- 16949: 8.5.5.2 Service agreement with customer

The automotive industry requires that the organization's QMS consider warranty management. It is essential that this be considered for inclusion in a facility's processes and interactions, given that it is a primary driver protecting the reputation of the both the OEM and the supplier and the financial viability of the project. Feedback from the field forms the basis for potential risk identification and mitigation early—before new field issues develop.

- 16949: 10.2.5 Warranty management

Appendix A: Leadership of top management

An organization's leadership sets the stage for how the entire business operates, including the QMS. A company's business objectives and strategy must be visionary to undertake present customer satisfaction, as well as future customer satisfaction. The two visions may or may not be compatible. However, they are essential for the survival of the organization. The goals for such mind-set transition must be in line with the general vision of the organization and meet the financial expectations of the stockholders. This new vision to satisfy future customers is not an easy task. In addition to the operating and financial challenges, leadership must also be prepared for expansion of production, as well as evaluating, sourcing, and developing the supply chain to accommodate the pending changes for the future.

Leadership required by IATF 16949 is proactive, involved, and can be considered on two levels: top management and throughout the supply chain. Top management is responsible for setting an example for the organization. While each organization's internal culture is unique, top management is responsible for providing the well-being of the whole organization. In the IATF 16949 standard, the terminology has been changed from "Management Responsibility" to "Leadership" to emphasize that management's role involves *leading*—through active engagement—to help ensure customer satisfaction. Top management looks outward at the potential market, identifies risks that threaten their ability to meet their commitments to their customer base, and allocates appropriate resources to mitigate those risks and put plans into action.

When setting goals, top management needs to consider the *whole picture* for the organization. The automotive scheme specifically requires that top management consider all *interested parties* when setting the strategic and operational objectives for the QMS. These requirements include both *internal* and *external* interested parties.

Corporate stewardship and leadership are the driving force behind an organization's QMS, as everyone looks to the leaders to *set* the tone within the organization. This sense of ownership is also needed to establish a single "voice" that directs the whole organization to mitigate the risks inherent to their business and fulfill the organization's commitments. Without this sort of overriding ownership, the resources and responsibilities needed to meet the needs of interested parties may fall into the "gaps" between the organization's operational units. Every part of the organization has a role to play in meeting customer requirements, and the role of leadership is to tie these parts together. The automotive scheme does not require the organization to identify a customer representative, but rather that it document roles and responsibilities throughout the organization. This mitigates the risk of important roles and responsibilities being lost due to employee attrition and bolsters organizational knowledge. Specifically, the IATF requirements for the identification and authenticity and requirements of leadership are:

- 16949: 5.1.1.3 Process owners
- 16949: 5.3.1 Organizational roles, responsibilities, and authorities—supplemental

Top management within an organization is clearly responsible for setting the direction and objectives of the organization and their QMS. Assuming management appropriately considered the needs of their interested parties, customer requirements would certainly be factored into the organization's planning operations. However, IATF 16949 adds some additional requirements to help ensure the organization has appropriate metrics and objectives to tie into their overall strategy. This helps to reinforce that setting objectives is not only about the very top level, but also about the processes that support the meeting of these objectives. The idea of *leadership* is not only tied to the top-level objectives for the facility or the business, but also applies to:

- Setting objectives at all levels.
- Leading the employees and mentoring them to contribute to the organization's goals. It's not only a question of how to do one's job, but to understand why one's doing it and how it fits into the big picture (organizational understanding).
- Leading and mentoring the supply chain to educate them about the organizational knowledge gained about the customer, including both the end-customer's explicit and implicit requirements.
  - 16949: 6.2.2.1 Quality objectives and planning to achieve them—supplemental

- 16949: 7.3.2 Employee motivation and empowerment
- 16949: 8.3.4.1 Monitoring

Leadership throughout the supply chain

It's essential for organizations to demonstrate leadership towards all suppliers under both their direct and indirect control. This differentiation of "direct" and "indirect" is necessary because sometimes an organization gets to choose their own suppliers and at other times the customer designates the suppliers the organization must use (a.k.a. "directed-buy"). Both of these kinds of suppliers are within the realm of responsibility of the organization's QMS and should be included when describing process sequence and interaction. In the automotive scheme, the organization is responsible for any and all outsourced operations.

- 16949: 4.4.1.1 Conformance of products and processes
- 16949: 8.4.1.3 Customer-directed sources (a.k.a. "directed-buy")

An organization is only as strong as the weakest link in its QMS, which includes the organization's supply chain. Original equipment manufacturers (OEMs) and some of the upper-tier suppliers may have very lengthy supply chains. Aligning and developing the supply chain around the needs of all interested parties is paramount to the organization's success. The IATF 16949 standard includes many requirements related to suppliers, but here we will focus on those that relate specifically to leadership. These requirements apply both to the products the organization chooses to outsource and to the *services* the organization contracts out.

- 16949: 8.4.1.1 General—supplemental

Selecting an adequate supplier for outsourcing is time-consuming, and a lot of resources are needed to develop suppliers for fruitful long-term relationships. The organization's top leadership needs to set the standard and manage their outsourced processes to align with the ultimate customer's requirements. The standard has a specific tactical methodology to implement this, but at a high level it focuses on the requirement to *develop the supplier and* their *QMS in relation to their performance.* This approach helps raise a supplier from where they are today to where the automotive industry needs them to be to provide continuing value.

Further development of the supplier occurs via two-way interaction between the supplier, who is responsible for meeting the requirements and working towards continual improvement, and the customer, who is required to adequately express what it really needs from the organization. This mutual exchange of ideas is refined during the "where known" requirements for product development, but at a top level requires leadership to make *intentional* requests to their supplier base and to communicate clear and unambiguous expectations.

- 16949: 8.4.3.1 Information for external providers—supplemental.

Appendix B: IATF 16949

The IATF 16949 standard has replaced the ISO/TS 16949 specification—effective September 14, 2018. Overall, the new standard has enhanced the automotive QMS, which further increases the value and credibility of certification. Furthermore, it has integrated many common CSR requirements into IATF 16949, and it has strengthened the performance linkage among OEMs, suppliers (all tiers), CBs, and oversight offices. Specifically, the changes in the IATF 16949 fall into three categories: (1) modified, (2) new changes, and (3) carryover from the ISO/TS standard. They are listed here without comments.

Modified CHANGES

Section 4.3.1: Determining the scope of the quality management system—supplemental

- These requirements were originally included in ISO/TS 16949:2009 Sections 1.1 and 1.2. They have been moved to Section 4 within IATF 16949.
- The requirement relating to supporting functions was revised to ensure that supporting functions not only address the need to include support functions in the audit, but also to ensure that they are included in the scope of the QMS.
- In addition, any exclusion sought for design and development activities, now in Section 8.3, has to be preserved as documented information.

CHANGES Section 5.1.1.2: Process effectiveness and efficiency

- The requirement for a supplier/organization to review their processes to ensure effectiveness and efficiency was covered in ISO/TS 16949, Section 5.1.1.
- Based on survey feedback, the IATF strengthened the requirement to ensure that the results of process review activities will now be included in management review.

- Process review activities need to include evaluation methods and, as a result, implement improvements.
- The results of these steps would be an input to the management review process. Top management is thus performing a review of the process-specific reviews performed by the process owner.

CHANGES Section 5.3.1: Organizational roles, responsibilities, and authorities—supplemental

- This requirement was already part of ISO/TS 16949:2009. However, based on IATF survey feedback, the IATF adopted some modifications to the requirement to address the need to document assigned personnel responsibilities and authorities.
- Additionally, this clause now clarifies that the goal is not just to address customer requirements but also to meet customer requirements fully.
- Personnel involved in capacity analysis, logistics information, customer scorecards, and customer portals now also need to be assigned and documented, per the requirements in this section.

CHANGES Section 5.3.2: Responsibility and authority for product requirements and corrective actions

- Based on survey feedback, the IATF adopted enhancements to the requirement originally included in ISO/TS 16949 to explicitly make top management responsible for ensuring conformity to product requirements and that corrective actions are taken.
- IATF 16949 clarifies that there must be a process to inform those with the authority and responsibility for corrective action in order that they ensure non-conforming product is identified, contained, and not shipped to the customer.
- This implies that the assigned personnel must be always available to take prompt action to prevent release.

Section 6.1.2.1: Risk analysis

- The need to identify, analyze, and consider actual and potential risks was covered in various areas of ISO/TS 16949.
- The IATF adopted additional requirements for risk analysis recognizing the continual need to analyze and respond to risk and to have suppliers/organizations consider specific risks associated with the automotive industry.
- Organizations would need to periodically review lessons learned from product recalls, product audits, field returns and repairs,

complaints, scrap, and rework, and implement action plans in light of these lessons.
- The effectiveness of these actions should be evaluated, and actions integrated in to the organization's QMS.

CHANGES Section 6.1.2.2: Preventive action

- The IATF enhanced the requirement found in ISO/TS 16949 by integrating what is considered to be a best practice in the automotive industry.
- Organizations would need to implement a process to lessen the impact of negative effects of risk, appropriate to the severity of the potential issues.
- Such a process would include: identifying the risk of nonconformity recurrence, documenting lessons learned, identifying and reviewing similar processes where the nonconformity could occur, and applying lessons learned to prevent potential occurrence.

CHANGES Section 6.1.2.3: Contingency plans

- The expanded requirement ensures the organization defines and prepares contingency plans along with a notification process to the customer or other interested parties.
- Organizations would first take a systematic approach to identifying and evaluating risk for all manufacturing processes, giving particular attention to external risk.
- Contingency plans would be developed for any of the outlined disruption conditions—interruption of externally provided products, processes, and services, recurring natural disasters, fire, or infrastructure-related disruptions.
- Customer notification is a mandatory step in any contingency plan unless there is no risk of delivering nonconforming product or affecting on-time delivery.

CHANGES Section 6.2.2.1: Quality objectives and planning to achieve them—supplemental

- ISO/TS 16949 included the importance of addressing customer expectations in the note to Section 5.4.1.1. The IATF enhanced the requirement by requiring that it be done at all levels throughout the organization.
- In ensuring quality objectives meet customer requirements, these objectives need to consider customer targets.
- Personnel should be aware of, and committed to, achieving results that meet customer requirements.

- Quality objectives and related performance targets should be periodically reviewed for adequacy (at least annually).

CHANGES Section 7.1.3.1: Plant, facility, and equipment planning

- This updated section includes an increased focus on risk identification and risk mitigation, evaluating manufacturing feasibility, re-evaluation of changes in processes, and inclusion of on-site supplier activities.
- Many operational risks can be avoided by applying risk-based thinking during planning activities, which also extends to optimization of material flow and use of floor space to control non-conforming product.
- Capacity planning evaluation during manufacturing feasibility assessments must consider customer-contracted production rates and volumes, not only current order levels.

CHANGES Section 7.1.5.1.1: Measurement system analysis

- Records are now required for customer acceptance of alternative methods. The previous requirement to analyze variation in measurement results is now extended specifically to inspection equipment.
- IATF 16949 also clarifies that records of customer acceptance need to be retained along with results from alternative measurement system analysis.

CHANGES Section 7.1.5.2.1: Calibration/verification records

- This updated section helps ensure that customer requirements are met through enhanced calibration/verification record retention requirements, including software installed on employee-owned or customer-owned equipment.
- IATF 16949 clarifies that a documented process is required to manage calibration/verification records in order to provide evidence of conformity, and this includes any on-site supplier owned equipment.
- Inspection, measurement, and test equipment calibration/verification activities need to consider applicable internal, customer, legislative, and regulatory requirements in order to establish approval criteria.

CHANGES Section 7.1.5.3.2: External laboratory

- This updated section allows the organization to conduct second-party assessments of laboratory facilities but requires customer-approval of the assessment method.

- The clause also clarifies that internal laboratory requirements apply even when calibration is performed by the equipment manufacturer, and that use of calibration services may be subject to government regulatory confirmation.

CHANGES Section 7.2.1: Competence—supplemental

- This section adds a requirement of "awareness," which includes knowledge of an organization's (supplier's) quality policy, quality objectives, personnel contribution to the QMS, benefits of improved performance, and implications of not conforming with QMS requirements.
- It also further emphasizes the customer requirements for OJT (on-the-job training), not just quality requirements.
- Note that the use of the term "process" rather than "procedure" implies that these activities need to be managed (via the plan-do-check-act cycle), and not merely performed.

CHANGES Section 7.2.2: Competence—on-the-job training

- IATF 16949 enhances the emphasis of OJT and its importance in meeting customer requirements, including other interested parties.
- The process would consider any relevant interested party requirements as an input in determining the need for OJT, and then consider the level of education and complexity of the tasks in determining the method used.
- This training must also include contract or agency personnel and convey the consequences of nonconformity to customer requirements to all persons whose work affects quality.

CHANGES Section 7.2.3: Internal auditor competency

- This section features greatly enhanced requirements to the organization's internal auditor competency to ensure a more robust internal audit process.
- Organizations need to establish a documented process that considers the competencies required by this clause, take actions to address any deficiencies, assess the effectiveness of actions taken, and record a list of the approved auditors.
- The clause differentiates between QMS auditors, manufacturing process auditors, and product auditors, and clarifies the competence requirements for each type of audit.

CHANGES Section 7.3.1: Awareness—supplemental

- Includes additional requirements to ensure all employees are aware of their impact on the organization's (supplier's) product quality output, customer specific requirements, and risks involved for the customer with non-conforming product.

CHANGES Section 7.5.1.1: Quality management system documentation

- The IATF retained the quality manual requirement that was removed in ISO 9001:2015; however, the quality manual can be one main document or a series of multiple documents (hard copy or electronic).
- This section also requires that the organization's processes and interactions are documented as part of their QMS.
- The quality manual needs to document where in the organization's QMS customer-specific requirements are addressed.

CHANGES Section 7.5.3.2.1: Record retention

- This section now requires a record retention process that is defined and documented and includes the organization's record retention requirements.
- Specifically calls out production part approvals, tooling records, product and process design records, purchase orders, and contracts/amendments.
- If there is no customer or regulatory agency retention period requirements for these types of records, "the length of time that the product is active for production and service requirements, plus one calendar year" applies.

CHANGES Section 7.5.3.2.2: Engineering specifications

- Added an engineering specifications requirement that the process is documented and agreed with the customer.
- This section also clarifies product design changes and product realization process changes, and the alignment to related sections.
- If there are no other overriding customer agreements, reviews of engineering standards/specifications changes should be completed within ten working days of receipt of notification.

CHANGES Section 8.1.1: Operational planning and control—supplemental

- This section features enhanced detail to ensure key processes are included and considered when planning for product realization.

- The required topics include customer product requirements and technical specifications, logistics requirements, manufacturing feasibility, project planning, and acceptance criteria.
- The section also clarifies the "resources needed to achieve conformity" encompasses all aspects of the development process, not just the manufacturing process requirements.

CHANGES Section 8.2.1.1: Customer communication—supplemental

- Added a requirement that the communication language (written or verbal) must be agreed with the customer.
- This should be considered when determining the necessary competence for roles that require customer communication.

CHANGES Section 8.2.2.1: Determining the requirements for products and services—supplemental

- The IATF strengthened the standard by elevating Notes 2 and 3 of the former clause into requirements.
- This suggests current organizational knowledge regarding recycling, environmental impact, and product and manufacturing process characteristics should be standardized.
- This knowledge would be systematically reviewed and used when determining the requirements for the products and services to be offered to customers.

CHANGES Section 8.2.3.1.1: Review of the requirements for products and services—supplemental

- IATF 16949 strengthens this requirement by requiring the organization to retain a documented customer authorization for waivers of formal reviews for products and services.

CHANGES Section 8.2.3.1.3: Organization manufacturing feasibility

- Enhanced requirements for manufacturing feasibility analysis through the following changes: Requiring a multidisciplinary approach to analyze feasibility, considering all engineering and capacity requirements.
- Requiring this analysis for any new manufacturing or product technology, and for any changed manufacturing process or product design.
- The organization should validate their ability to make product specifications at the required rate. These should consider customer-specific requirements.

CHANGES Section 8.3.1.1: Design and development of products and services—supplemental

- Strengthened the standard by elevating the note in the former section to a requirement and added a requirement for documentation of the design and development process.
- As the concept of the design and development process in the automotive industry includes manufacturing design and development, the requirements from other parts in Section 8 should be considered complimentary in the context of manufacturing and product design and development.

CHANGES Section 8.3.2.1: Design and development planning—supplemental

- Clarifies when the multidisciplinary approach is to be used and who should be involved. Specifically, it must include all affected stakeholders within the organization and, as appropriate, its supply chain.
- Additional examples are provided of areas where such an approach may be used during design and development planning (including project management), and the note further clarifies that purchasing, supplier, and maintenance functions might be included as stakeholders.

CHANGES Section 8.3.3.1: Product design input

- This section expanded the minimum set of product design input requirements, emphasizing regulatory and software requirements.
- New and broadened requirements include: product specifications; boundary and interface requirements; consideration of design alternatives; assessment of risks and the organization's ability to mitigate/manage those risks; conformity targets for preservation, serviceability, health, safety, environmental, and development timing; statutory and regulatory requirements for the country of destination; and embedded software requirements.

CHANGES Section 8.3.3.2: Manufacturing process design input

- Expanded the list of manufacturing process design inputs, including: product design output data including special characteristics and targets for timing; manufacturing technology alternatives; new materials; product handling and ergonomic requirements; and design for manufacturing and design for assembly.

- This could include consideration of alternatives from innovation and benchmarking results, and new materials in the supply chain that could be used to improve the manufacturing process capacity.
- This section also further strengthened the requirements by transforming the former note regarding error-proofing methods into a requirement.

CHANGES Section 8.3.3.3: Special characteristics

- Identify the source of special characteristics and including risk analysis to be performed by the customer or the organization.
- Expands the list of sources used to identify special characteristics, along with the requirements related to those special characteristics.
- Special characteristics need to be marked in all applicable cascaded quality planning documents; monitoring strategies should focus on reducing variation, which is typically done using statistical techniques.
- The organization must also consider customer-specific requirements for approvals and use of certain definitions and symbols, including submission of the symbol conversion table, if applicable and required.

CHANGES Section 8.3.4.1: Monitoring

- The requirement clarifies that measurements apply at specified stages during the design and development of both products and services, and that reporting must occur as required by the customer.
- This could include, for example, the periodic update of customer APQP schedule milestones, gate reviews, and open issues lists related to development activities.

CHANGES Section 8.3.4.2: Design and development validation

- This section features a strengthening of the requirements for design and development validation and also added embedded software.
- Customer specific requirements (CSRs), industry, and governmental agency-issued regulatory standards need to be considered when planning and performing design and development activities.

CHANGES Section 8.3.4.3: Prototype program

- The changes in this section strengthen the standard by focusing the organization on the QMS for managing outsourced products and services.

- Regardless of whether the work is performed by the organization or by an outsourced process, the prototype program and control plan are part of the scope of the QMS.
- This type of control should be considered a support process and be integrated into the design and development process.

CHANGES Section 8.3.4.4: Product approval process

- These changes clarify approval requirements, with an emphasis on outsourced products and/or services and record retention required.
- The activities should be managed (with an effectiveness review and improvement actions applied) and not just performed.
- A part approval process for externally provided products and services needs to be performed prior to final product submission to customers.
- Product approval must be obtained when the customer requires it, and records retained.

CHANGES Section 8.3.5.1: Design and development outputs—supplemental

- Product design output additions include recognition of the use of 3D models, and inclusion of service parts and packaging.
- IATF 16949 clarifies that it requires product design error proofing methods, such as DFSS, DFM/A, and FTA. The application of GD&T tolerancing and positioning systems allows organizations to specify dimensions and related tolerances based on functionality relationships.
- Outputs include repair and serviceability instructions and service parts requirements that will be used by approved maintenance organizations.

CHANGES Section 8.3.5.2: Manufacturing process design output

- Changes in this section strengthened verification requirements, process input variables, capacity analysis, maintenance plans and correction of process nonconformities.
- Clarifies that the process approach methodology of verifying outputs against inputs applies to the manufacturing design process.
- The list of manufacturing design outputs is also expanded.

CHANGES Section 8.3.6.1: Design and development changes—supplemental

- This section strengthens the requirement for change validation and approval prior to implementation, and also added embedded software.

- Design changes after initial product approval imply that products, components, and materials need to be evaluated and validated prior to production implementation.
- This validation needs to be done by the organization and the customer when there is a CSR.
- For products with embedded software, the change record needs to document the revision level of the software and hardware to help assure that product configuration is managed appropriately.

CHANGES Section 8.4.1.1: General—supplemental (under Control of externally provided processes, products and services)

- The former note about purchased products was broadened and elevated into a requirement.
- It now clarifies that all the requirements of Section 8.4 apply to subassembly, sequencing, sorting, rework, and calibration services.

CHANGES Section 8.4.1.2: Supplier selection process

- While ISO/TS 16949:2009 did address supplier selection in the ISO 9001:2008 boxed text via the Purchasing Process (see Section 7.4.1), the supplier selection process was not as detailed.
- This section now specifically calls out supplier selection process criteria, in addition to clarifying that it is a full process.
- The assessment used to select suppliers needs to be extended beyond typical QMS audits and include aspects such as: risk to product conformity and uninterrupted supply of the organization's product to their customers, etc.

CHANGES Section 8.4.1.3: Customer-directed sources (also known as "directed-buy")

- This section features a clarification of the organization's responsibilities for customer directed sources, even for customer directed-buy suppliers.
- Unless otherwise defined by contract, all requirements of IATF 16949 Section 8.4 apply in this situation, except requirements related to the selection of the supplier itself.

CHANGES Section 8.4.2.1: Type and extent of control—supplemental

- The changes in this section further strengthened the requirement for control of outsourced processes, including the assessment of risk.

- Internal and customer requirements are inputs that need to be considered during the development of methods to control externally provided products, processes, and services.
- Type and control need to be consistent with supplier performance and an assessment of product, material, or service risk.
- This implies a constant monitoring of performance and assessment of risk based on the established criteria, triggering the actions to escalate (increase) or reduce the types and extent of control.

CHANGES Section 8.4.2.2: Statutory and regulatory requirements

- The updates clarify the applicability of statutory and regulatory requirements and strengthen the requirements.
- Identification of applicable statutory and regulatory requirements needs to consider the country of receipt, shipment, and delivery.
- When special controls are required, the organization must implement these requirements and cascade those requirements down to their suppliers.

CHANGES Section 8.4.2.3: Supplier QMS development

- This section provides a method to strengthen ISO 9001 certification, aligns with CSRs, and clarifies the acceptable third-party certification bodies (which shall be recognized by the IATF).
- Instead of requiring organizations to simply "develop" the supplier QMS, this section outlines a progressive approach that goes from compliance to ISO 9001 via second-party audits all the way through certification to IATF 16949 through third-party certification.

CHANGES Section 8.4.2.4: Supplier monitoring

- Organizations should continuously review inputs and introduce improvement actions regarding supplier-monitoring data, as needed.
- Documented and non-documented yard holds and stop ships should be considered customer disruptions, and the number of premium freight occurrences need to be monitored.
- Performance indicators provided by the customer and from service need to be included within the organization's supplier monitoring process.

CHANGES Section 8.4.2.5: Supplier development

- This section adds an emphasis on performance-based supplier development activities.

- Supplier monitoring process should be considered an input to supplier development activities. These development activities should consider both short-term and long-term goals.
- Short-term efforts would generally focus on supplier products and would require defining suitable methods to assure the quality of purchased product from each supplier.
- Long-term efforts would generally focus on supplier QMS and manufacturing processes as a whole, and consider audits, training, and enhancement efforts that implement and enhance quality assurance agreements between suppliers and the organization, and further reduce risk.

CHANGES Section 8.4.3.1: Information for external providers—supplemental

- The organization is required to provide key information to their supply chain through this new requirement.
- This information includes all applicable statutory and regulatory requirements and special product and process characteristics.

CHANGES Section 8.5.1.1: Control plan

- This section strengthened the control plan requirements and aligned IATF OEM CSRs into the IATF 16949 standard. It also elevated a note regarding customer approval to a requirement and strengthened the control plan review and update criteria and linked to the PFMEA updates.
- Control plans are needed for the relevant manufacturing site and all product supplied, and not just for the final product or final assembly line, as an example.
- Although family control plans are acceptable for bulk material and similar parts using a common manufacturing process, care should be given to identify the degree of difference that is acceptable to apply this common control.

CHANGES Section 8.5.1.2: Standardized work—operator instructions and visual standards

- Through this section, IATF 16949 strengthens the requirements for standardized work, including the requirement to address specific language needs.
- Standardized work documents need to be clearly understood by the organization's operators and should include all applicable quality, safety, and other aspects necessary to consistently perform each manufacturing operation.

CHANGES Section 8.5.1.3: Verification of job setups

- The changes in this section elevate a NOTE to a requirement and clarify record retention.
- Clarify that the organization shall verify job changes that require a new setup; maintain documented information for setup personnel; perform first-off/last-off part validation, as applicable, including retention and comparison; and retain records of process and product approval following these validation actions.

CHANGES Section 8.5.1.5: Total productive maintenance

- Strengthens the requirement for equipment maintenance and overall proactive management of the Total Productive Maintenance (TPM).
- TPM is a system for maintaining and improving the integrity of production and quality systems through machines, equipment, processes, and employees that add value to the manufacturing process. TPM should be fully integrated within the manufacturing processes and any necessary support processes.

CHANGES Section 8.5.1.6: Management of production tooling and manufacturing, test, inspection tooling, and equipment

- IATF 16949 features strengthened tooling and equipment marking and tracking requirements.
- This requirement extends the scope to production and service materials and for bulk materials, as applicable, and clarifies that requirements apply whether tooling is owned by the organization or by the customer.
- The updates clarify that the system for production tooling management must include tool design modification documentation and tool identification information.
- Customer-owned tools and equipment need to be permanently marked in a visible location.

CHANGES Section 8.5.1.7: Production scheduling

- This section emphasized the importance of planning information and integrated IATF OEM customer lessons learned.
- Ensure that customer orders/demands will be achieved.
- This suggests the organization needs a robust feasibility review process regarding production scheduling. The production scheduling activities also need to include all relevant planning information as inputs to their feasibility review and make any necessary adjustments.

CHANGES Section 8.5.2.1: Identification and traceability—supplemental

- Strengthened the requirements for traceability to support industry lessons learned related to field issues.
- Requirement of clear start and stop points for product received by the customer is aligned with the definition of traceability in ISO 9000:2015.

CHANGES Section 8.5.4.1: Preservation—supplemental

- Adds specificity to preservation controls and includes application to internal and/or external providers.
- Preservation activities are expanded in two ways: first, activities that are considered preservation controls, and second, locations where preservation controls apply.
- Preservation controls include the preservation of identification during the product shelf life; a contamination control program appropriate to identified risks; design and development of robust packaging and storage areas; adequate transmission and transportation considerations; and measures to protect product integrity.

CHANGES Section 8.5.5.1: Feedback of information from service

- Requirements for this section feature an expanded scope to include material handling and logistics.
- The new second note also clarifies that "service concerns" should include the results of field failure test analysis where applicable.
- The intent of this addition is to ensure that the organization is aware of nonconformities that occur outside of its organization.

CHANGES Section 8.5.5.2: Service agreement with customer

- This section clarifies that service centers need to comply with all applicable requirements when there is a service agreement with the customer.

CHANGES Section 8.5.6.1: Control of changes—supplemental

- IATF 16949 strengthens the control of changes requirements in the standard to align with existing IATF OEM requirements.
- The changes clarify that "any change" includes those caused by the organization and/or the customer, in addition to those by any supplier.

- The process to control and react to changes needs to include risk analysis and to retain records of verification and validation.
- FMEAs should be reviewed for any manufacturing or product changes, prior to implementation. Production trial run activities should be planned based on the risk and complexity of the changes.

CHANGES Section 8.6.1: Release of products and services—supplemental

- While ISO/TS 16949:2009 did mention product and delivery of service in the ISO 9001:2008 boxed text via the Monitoring and Measurement of Product section (see Section 7.4.1), the product and delivery of service process is further detailed in IATF 16949.
- These updates strengthen the standard to ensure process controls align with the control plan.
- To achieve coherence between the control plan and the planned arrangements to verify product and service conformity, the organization should conduct a regular control plan audit that compares the current approval status of the product and process with the actual controls applied in the manufacturing process.

CHANGES Section 8.6.2: Layout inspection and functional testing

- An added note clarifies that frequency of layout inspections is determined by the customer.

CHANGES Section 8.6.3: Appearance items

- This section requires organizations to provide masters for haptic (sense of touching) technology, as appropriate. Haptic technology recreates the sense of touch by applying forces, vibrations, or motions to the user.

CHANGES Section 8.6.4: Verification and acceptance of conformity of externally provided products and services

- Changes in this section align with ISO 9001:2015 terminology and clarify the source of statistical data as that provided by the supplier to the organization.

CHANGES Section 8.6.5: Statutory and regulatory conformity

- Strengthens the standard for statutory and regulatory conformity to require evidence of compliance.
- "Prior to release" means that the organization should implement some process and/or agreements with its suppliers requiring

sufficient prevention and detection controls to ensure that products meet all applicable statutory, regulatory, and other requirements.
- These requirements must consider both the countries where products are manufactured and the destination countries.

CHANGES Section 8.7.1.1: Customer authorization for concession

- Changes in this section are for the alignment of terminology, and the clarification of concessions applied to rework of nonconforming product and subcomponent reuse.
- The changes clarify that the organization must obtain customer authorization prior to further processing for "use as is" and rework disposition of nonconforming products, and sub-component reuse must be clearly communicated to the customer.
- Appropriate internal verification and validation activities of any rework or reuse of sub-components should be approved prior to customer submission.

CHANGES Section 8.7.1.2: Control of nonconforming product—customer specified process

- This section ensures customer-controlled shipping requirements are followed, and that these customer-specific requirements are integrated into the organization's internal activities for the control of nonconforming product.

CHANGES Section 8.7.1.3: Control of suspect product

- The updates in this section augment the requirements for control of suspect product by ensuring containment training is implemented.
- Appropriate training should consider, for example, awareness about special characteristics, CSRs related to nonconforming product control, product safety, escalation processes, storage areas, and related roles.

CHANGES Section 8.7.1.4: Control of reworked product

- This update increases the scope of control of reworked product requirements to include: customer approval, risk assessment, rework confirmation, traceability, and retention of documented information.
- The risk analysis and customer approval requirements are interrelated; FMEAs should identify and address risks related to each possible rework of the characteristics stated in the control plan.

CHANGES Section 8.7.1.5: Control of repaired product

- The changes in this section clarify the requirement and the need for follow-up with detailed information for reworked product.

CHANGES Section 8.7.1.7: Nonconforming product disposition

- Strengthen the requirement of disposition of nonconforming product by clarifying that organizations must also have a documented process for disposition of nonconforming product not subject to rework or repair.
- Planned activities need to be managed and the results considered to improve this process.
- Contamination control practices should be applied to avoid any risk of unintended use of this type of nonconforming products.
- Customer approval is required before nonconforming products in this category can be diverted for service or any other use.

CHANGES Section 9.1.1.1: Monitoring and measurement of manufacturing processes

- Clarifies the requirement for targeting process effectiveness and efficiency (not just *having* a process, but monitoring it).
- Further ensures that organizations support the manufacturing process through defined roles, responsibilities, and effective escalation processes to drive process capability and stability.
- The note clarifies that it may not be possible or feasible to measure product or manufacturing process characteristics through process capability assessments. In such cases, a rate or index of lot conformity may be acceptable.

CHANGES Section 9.1.1.2: Identification of statistical tools

- Requirements for the identification of statistical tools feature clarifications regarding the documented deployment of the use of statistical tools from DFMEA, PFMEA, and the APQP (or equivalent) process.
- The tool chosen in the APQP (or equivalent) process must be included in design/process risk analysis and the control plan.

CHANGES Section 9.1.1.3: Application of statistical concepts

- This section features a clarification regarding requirements for those involved in capturing and analyzing data; previously, this was driven across all employees regardless of relevance.

- These concepts should be included in the competencies required for "employees involved in the collection, analysis, and management of statistical data."

CHANGES Section 9.1.3.1: Prioritization

- The emphasis of the requirement changed from the ISO/TS 16949 standard's "Analysis of data" to the prioritization of actions based on performance and risk management.
- Actions to improve customer satisfaction need to take precedence as the organization considers trends and drives towards improvement.

CHANGES Section 9.2.2.1: Internal audit program

- Strengthened the need to drive a risk-based approach to the development and deployment of an organization wide internal audit program.
- Internal audit activities are considered a process that requires a clear definition of expected inputs, planned activities, intended outputs, and monitored performance.
- The process needs to identify and evaluate the level of risk related to each QMS process, internal and external performance trends, and process criticality.
- Then, the process would need to continuously monitor this information to trigger special internal audits and/or to plan periodic internal audits.

CHANGES Section 9.2.2.2: Quality management system audit

- Strengthen the quality management system audit and the use of process approach, which further drives process improvements organization wide.
- The audit program is continuously monitoring information that could trigger the need for an unplanned internal audit.
- The use of the automotive process approach, including risk-based thinking, needs to be applied during the audit.
- The internal audit must also sample customer-specific QMS requirements for effective implementation.

CHANGES Section 9.2.2.3: Manufacturing process audit

- Strengthens the formal approaches to ensure organizations achieve the benefits of effective manufacturing process audits.

- Shift handover should be considered a significant process event; internal auditors should look for objective evidence of an effective process to communicate and address relevant information.
- The audit must also evaluate the effective implementation of the process risk analysis, control plan, and associated documents.

CHANGES Section 9.2.2.4: Product audit

- The strengthened product audit requirements now require the use of customer-specified approaches, when applicable.
- If not applicable, the organization shall define their process.

CHANGES Section 9.3.1.1: Management review—supplemental

- Strengthens management review requirements to include an assessment of risk and compliance with customer requirements.
- The one-year frequency is a minimum, as the process is driven by the continuous assessment of the risks related to internal and external changes and performance related issues.
- As changes and issues increase, the frequency of management review activities should increase in turn, preserving the minimum of at least an annual review.

CHANGES Section 9.3.2.1: Management review inputs—supplemental

- Enhanced details for management review input requirements, including those related to cost of poor quality, effectiveness, efficiency, conformance, feasibility assessments, customer satisfaction, performance against maintenance objectives, warranty performance, review of customer scorecards, and the identification of potential field failures through risk analysis.
- The above should be considered the minimum information to be covered during management review; a monitoring system should be in place, with criteria that trigger special unplanned management review activities.

CHANGES Section 9.3.3.1: Management review outputs–supplemental

- Enhanced section ensures action is taken where customer requirements are not achieved and supports the continual analysis of process performance and risk.
- Even though process owners should address customer performance issues related to the processes they manage, this requirement gives

top management the clear and ultimate responsibility to address customer performance issues and ensure the effectiveness of corrective actions.

CHANGES Section 10.2.3: Problem solving

- Updates to this section are to facilitate the consolidation of IATF OEM customer specific minimum requirements.
- The organization's defined process(es) for problem solving must consider: various types and scales of problems; control of nonconforming output; systemic corrective action and verification of effectiveness; and review/updates to documented information.
- In addition, CSRs related to nonconformity and corrective action need to be used and integrated within the internal corrective action process.

CHANGES Section 10.2.4: Error-proofing

- This section, which previously only mentioned the use of error proofing methods in corrective action, includes new requirements to strengthen the approach to error proofing and consolidate CSRs.
- The organization needs a process that both identifies the need or opportunity for an error-proofing device/method, and designs and implements the device/method.
- The FMEA would document whether the method impacts occurrence (prevention control) or impacts detection (detection control).
- The control plan needs to include the test frequency of the error-proofing devices, and records must be maintained for the performance of these tests.

CHANGES Section 10.2.6: Customer complaints and field failure test analysis

- Includes a new requirement regarding embedded software and identification of preferred approaches.
- The organization's analysis is extended beyond parts to customer complaints and field failures themselves, and the results must be communicated to the customer and also within the organization.

CHANGES Section 10.3.1: Continual improvement–supplemental

- Changes in this section clarify the minimum process requirements for continual improvement: identification of methods, information and data; an improvement action plan that reduces variation and waste; and risk analysis (such as FMEA).

- Use of TPM, Lean, Six Sigma, and other manufacturing excellence programs or methodologies should follow a structured approach that continuously identifies and addresses opportunities for improvement.

New CHANGES

Section 4.3.2: Customer-specific requirements

- Although the need to fulfill and satisfy CSRs was already mentioned throughout the whole ISO/TS 16949 document, in IATF 16949 this requirement specifically addresses the need to evaluate the customer specific requirements and include them where applicable in the organization's quality management system.
- This means that the supplier would need some sort of process to evaluate each of their customer's CSRs and determine exactly how (and where) it applies to their organization's QMS, as applicable.

CHANGES Section 4.4.1.1: Conformance of products and processes

- This requirement was adopted based on IATF survey feedback received.
- It ensures two things:
  - that the supplier (organization) is responsible for the conformity of outsourced processes, and
  - that all products and processes meet all applicable requirements and expectations of all interested parties.
- To ensure conformance of all products and processes, the organization would need to take a proactive approach to assess and address risks, and not rely only on inspection.

CHANGES Section 4.4.1.2: Product safety

- New section with enhanced requirements that address current and emerging issues the automotive industry is facing related to product and process safety.
- Organizations (suppliers) are required to have documented processes to manage product-safety related products and processes.

CHANGES Section 4.4.1.2: Product safety

- This section includes identification of statutory requirements; identifying and controlling product-safety related characteristics both during design and at point of manufacture; defining responsibilities, escalation processes, reaction plans, and the necessary flow of information including top management and customers; receiving special approvals for FMEAs and Control Plans; product traceability measures; and cascading of requirements throughout the supply chain.

CHANGES Section 5.1.1.1: Corporate responsibility

- ISO 9001:2015 expanded the ISO 9001:2009 concept of management responsibility into a set of leadership behaviors to ensure an effective QMS.
- IATF 16949 includes the requirement for an anti-bribery policy, an employee code of conduct, and an ethics escalation policy to address increasing market and governmental expectations for improved integrity in social and environmental matters in the automotive industry.
- This implies responsibility and empowerment at all levels and functions of the supplier/organization to follow an ethical approach and report any observed unethical behavior without fear of reprisal.

CHANGES Section 5.1.1.3: Process owners

- ISO/TS 16949:2009 addresses management responsibility and authority, but it does not explicitly mention that management ensure process owners understand their role and are competent.
- The IATF adopted this new requirement to ensure that management understands this expectation, by specifically identifying these process owners and ensuring they can perform their assigned roles.
- This requirement recognizes that process owners have the authority and responsibility for activities and results for the processes they manage.

CHANGES Section 7.2.4: Second-party auditor competency

- This new section outlines requirements for second-party auditors, ensuring they are properly qualified to conduct those types of audits, with CSRs being a main focus.
- The same core competencies that apply to internal auditors should, at a minimum, also apply to second-party auditors.

CHANGES Section 8.3.2.3: Development of products with embedded software

- This new clause adds requirements for organization-responsible embedded software development and software development capability self-assessments.

- Organizations must use a process for quality assurance for products with internally developed embedded software and have an appropriate assessment methodology for their software development process.
- The software development process must also be included within the scope of the internal audit program; the internal auditor should be able to understand and assess the effectiveness of the software development assessment methodology chosen by the organization.

CHANGES Section 8.4.2.3.1: Automotive product-related software or automotive products with embedded software

- This new section added requirements for software development assessment methodology.
- These requirements align to those presented within Section 8.3 but are now cascaded down to suppliers.

CHANGES Section 8.4.2.4.1: Second-party audits

- This new section aligns customer-specific requirements into the IATF 16949 standard.
- Second-party audits should consider issues relevant to the organization beyond simply the maturity of their QMS development.
- Examples of situations that could trigger a second-party audit include: input from supplier performance indicators; risk assessment results and follow-up of open issues from process and product audits; and new development launch readiness.
- The organization's criteria for determining the need, type, frequency, and scope of second-party audits must be based on a risk analysis.

CHANGES Section 8.5.1.4: Verification after shutdown

- Defines a new requirement for verification after shutdown, integrating industry lessons learned and/or best practices. The necessary actions after the shutdown period should be anticipated in the PFMEA, control plans, and maintenance instructions, as appropriate.
- A multidisciplinary approach should be used to identify any additional actions needed to address unexpected shutdown events.

CHANGES Section 8.5.6.1.1: Temporary change of process controls

- This new requirement for temporary control of process changes addresses issues experienced by the IATF OEM customers.

- The organization must identify, document, and maintain a list of process controls that includes both the primary process control (example: automated nut driver) and the approved back-up or alternate methods (example: manual torque wrench). The list must be updated regularly to reflect the current and approved process controls.
- The use of alternative control methods is considered a process; therefore, the organization is expected to manage these activities appropriately.

CHANGES Section 8.7.1.6: Customer notification

- This new section features a new automotive requirement to address modifications in ISO 9001 requirements and address customer issues for IATF OEM concerns. While customer notification is mentioned twice in ISO/TS 16949:2009 (see Sections 7.4.3.2 and 8.2.1.1), it did not address customer notification in a standalone section.
- The organization is required to immediately notify the customer if they ship nonconforming product and follow up with detailed documentation.

CHANGES Section 10.2.5: Warranty management systems

- This is a new requirement based on the increasing importance of warranty management and consolidates IATF OEM customer specific requirements.
- The warranty management process should address and integrate all applicable customer-specific requirements, and warranty part analysis procedures to validate no trouble found (NTF) decisions should be agreed by the customer, when applicable.

Carryover CHANGES

Section 7.1.4.1: Environment for the operation of processes—supplemental

- This requirement for an organization to "maintain its premises in a state of order, cleanliness, and repair" was preserved from ISO/TS 16949 and transferred to IATF 16949.

CHANGES Section 7.3.2: Employee motivation and empowerment

- This section did not substantially change, but now requires "maintain[ing] a documented process(es)" for employee motivation and empowerment, instead of simply "having a process."

CHANGES Section 8.1.2: Confidentiality

- Only a minor edit to clarify confidentiality "includes" related product information, instead of using the word "and." There is no change in intent.

CHANGES Section 8.2.3.1.2: Customer-designated special characteristics

- This section changes the action from "demonstrate conformity" to "conform," and clarifies that it refers to "approval documentation," rather than just "documentation."
- There is no change in intent.

CHANGES Section 8.3.2.2: Product design skills

- This section adds a note as an example of a product design skill-set. There is no change in intent.

CHANGES Section 8.6.6: Acceptance criteria

- This section clarifies "where required" to be "where appropriate or required," and updates the clause reference to align with the new structure.
- There is no major change in the intent of this section.

Appendix C: FMEA forms

Table C.1 Classic DFMEA form

POTENTIAL
FAILURE MODE AND EFFECTS ANALYSIS
(DESIGN FMEA)

_____ System

_____ Subsystem

_____ Component: ______________

Model Year(s)/Program(s) ______________

Design Responsibility ______________

Key Date ______________

FMEA Number ______________

Page ________ of ________

Prepared By ______________

FMEA Date (Orig.) ______________

Core Team ______________

Item / Function	Requirements	Potential Failure Mode	Potential Effect(s) of Failure	Severity	Classification	Potential Cause(s) of Failure	Occurrence	Current Design Controls Prevention	Current Design Controls Detection	Detection	RPN	Recommended Action	Responsibility & Target Completion Date	Action Results				
														Actions Taken & Effective Date	Severity	Occurance	Detection	RPN

Table C.2 Classic PFMEA form

POTENTIAL
FAILURE MODE AND EFFECTS ANALYSIS
(PROCESS FMEA)

FMEA Number________

Page ______ of ______

Item: ________ Process Responsibility ________ Prepared By: ________

Model Year(s)/Program(s) ________ Key Date ________ FMEA Date (Orig.) ________

Core Team ________

Process Step / Function	Requirements	Potential Failure Mode	Potential Effect(s) of Failure	Severity	Classification	Potential Cause(s) of Failure	Occurrence	Current Process Controls Prevention	Current Process Controls Detection	Detection	RPN	Recommended Action	Responsibility & Target Completion Date	Action Results				
														Actions Taken & Effective Date	Severity	Occurance	Detection	RPN

Table C.3 VDA generic FMEA form (Notice that there is no column for Classification, RPN, and corrective action is separated into preventive and detection columns)

Company Name:	Subject:	DFMEA ID number:
Engineering Location:	FMEA start date:	Design responsibility:
Customer Name:	FMEA revision date:	Security classification:
Model/Year/ Platform:	FMEA due date:	
FMEA Team:		

Structured Analysis			**Function Analysis**			**Failure Analysis**				**Risk Analysis**					
1. System item	2. System element or interface	3. Component element (item/interface	1. Function of system and requirement or intended output	2. Function of system element and intended performance output	3. Function of component element and requirement or intended output or characteristic	1. Failure effects (FE)	2. Severity (S) of FE	3. Failure mode (FM)	4. Failure cause (FC)	1. Current prevention control (PC) of FC	2. Occurrence (O) of FC	3. Current detection control (DC) of FC or FM	4. Detection (D) of FC or FM	5. Action priority (AP)	6. Filter code (optional)

Optimization										
Prevention action	Detection action	Responsible person	Target completion date	Status: (untouched, under consideration, in progress, completed, discarded	Action taken with pointer to evidence	Completion date	Severity (S)	Occurrence (O)	Detection (D)	Action priority (AP)

Appendix D: Failure mode avoidance (FMA)

Overview

In any endeavor to minimize or eliminate failures through any type of FMEA—especially DFMEA—avoidance must be the goal. However, in order to understand avoidance, we must review three definitions.

1. *Failure mode* is the manner in which a component, subsystem, or system could potentially fail to meet or deliver the intended function(s) or expectations.
2. *Failure mode avoidance* (*FMA*) is a systematic approach to the execution of engineering disciplines to deliver customer expectations while avoiding product and process failure modes through maximizing robustness and reducing or eliminating mistakes. FMA success is measured through successful engineering first and quality metrics second. FMA processes and tools are essential parts of the engineering process.
3. *Countermeasures* are engineering actions taken to eliminate, avoid, prevent, or compensate for a failure mode.

To carry out FMA there are three principles that are integrated into the mind setting of engineering, i.e., product development:

1. Failures are more than "things that break."
2. Better to prevent failure modes rather than detecting them. The old paradigm of "if it is not broken do not fix it" is not good enough. The new motto should be: "if it is not broken, improve it!"
3. Failures only need to be found and fixed once—this should be at the earliest opportunity. However, this is not enough. After correction, prevention should be undertaken so that the failure will not recur. This means that effective countermeasures are taken in the development phase when the failure was created.

To address these fundamental principles, we must deal with reliability. Specifically, the framework that deals with the two types of failure mode: (a) hard—something breaks; and (b) soft—performance degrades over time. In addition, we must recognize two distinct root causes, which are (a) mistakes—failures to take an action which is known to avoid a failure mode (the detection of mistakes, and our ability to avoid them, can be easily tracked, in a QOS)—and (b) lack of robustness—sensitivity to noise factors (the detection of robustness failures, and development of appropriate counter-measures requires rigorous technical analysis).

In the case of mistakes, the following must be considered:

Stages

- A mistake is not adopting a *known* countermeasure for a known failure mode.
- The countermeasure for mistakes (i.e., avoiding mistakes) is primarily a matter of vigilance.
- In large organizations, individual vigilance can be improved by giving people the chance to make the same mistake twice and supported by, for example, continued research into better practices, design guidelines, health charts, and standards with an enforced deviation process (deviations should be exceptional), and design reviews within peer groups.
- An engineering QOS system to monitor the occurrence of mistakes, so that corrections can be implemented quickly. It pays to be paranoid—in complicated and large organizations, mistakes are the entropic state. Let us all remember Henry Ford's words: "Trust but verify."
- Always remember that reliability is failure avoidance.

In the case of sensitivity, the following must be considered:

Some failure modes are caused by the sensitivity of the engineering function (physics, geometry, and properties of materials) to the noise factors. The five types of noises can be reduced to two categories. They are shown in Table D.1.

So, the very basic questions for an effective FMA are:

1. What is the system?
2. What does it do?
3. What failures have the customers seen either from similar or current designs?
4. What countermeasures avoid these failures?
5. Are these countermeasures effective?

Table D.1 Categories of noises

Types of noises	
Capacity noises (a.k.a. inner noises)	*Demand noises (a.k.a. outer noises)*
Production variation due to rate	Customer duty cycles
Wear out and drift over time	Environment
	Component interactions

The countermeasures for these failures are to improve robustness by increasing the *distance from the failure mode* (i.e., minimizing overlap between demand and capacity—stress and strength). Specifically, some options are: (a) reduce magnitude of demand, (b) reduce dispersion in capacity, (c) increase the nominal capacity, and (d) develop the countermeasures to manage the noise factors.
Robustness failures occur when the demand placed on the design exceeds the capacity for some units.

Key FMA activities

Quality history

Its function is to understand feedback, validate assumptions, learn from failures (TGW), successes (TGR), R/1000, and select appropriate and applicable tools. These must be discussed with appropriate engineers so that the feedback is legitimate and most importantly, so they understand how failure is reported. To this end, it is imperative that root cause and escape point must be discussed, i.e., where are the gaps in the design process?

Hopefully, understanding the quality history, wrong designs can be reviewed and applicable tests and controls can be instituted to existing and or future designs. That review should set priorities for programs and FMA in areas such as (a) select priority systems, (b) identify issues to fix, and (c) plan FMA tool usage.

FM plan

Why

1. To know what's gone wrong in the past.
2. Cause and corrective action.
3. Ensure we do not repeat the issue.
4. Ensure we have a test to detect the issue if does occur again.

Mindset/Best Practice

- Breadth and depth to identify issues.
- Multiple data sources used—corporate databases (BSAQ, AWS, Global 8D, Six Sigma, PTS, etc.) queried, analyzed, and prioritized.

- Review cross products—surrogate designs—similar to the proposed design and application, not just the prior design on the product.
- Campaign and recall analysis complete.
- Quality history over useful life—high TIS not just low TIS issues.
- Customer needs and wants are refined through analysis—decode issues from indicators. Good tools are the usage of the Kano model and QFD—quality function deployment.
- Explain failure—detailed root cause analysis—what was the engineering mechanism? Was the issue a failure to specify, test, or produce?
- PD process escape point understood—when in the product development process did failure occur?
- Countermeasures—which design features and rules have changed since the issue?
- Test improvements—what test procedure changes have been improved?

Boundary diagrams

Why

- Scope of the analysis
- Visualize the design
- Understand the system complexity and interactions
- Become aware of significant neighboring systems

Mindset

- Parts
- All parts necessary for the system to function shown
- Nearby parts also considered
- Energies considered—heat, electrical, chemical, mechanical, light, sound
- Hierarchy—always one higher and one lower subsystem, and the interface with the customer included
- Interfaces—every person, place, and thing should be considered

Interface

Why

- Add more rigor to analysis of interfaces
- Specifically detail the precise nature of interfaces and how they may drive failures

Mindset

- Detailed and structured analysis
- Full detail on every interface

- Parts
- All commodities necessary for the system to operate are shown
- Neighboring systems and influences considered
- Energies assessed—heat, electrical, chemical, mechanical, light, sound

Best Practice

- All system components and interfacing components shown
- All energy flows/signal or force transfers identified
- Interfaces quantified in engineering-measurable terms
- Program-specific values specific to technologies/hardware
- Owner defined for cross boundary interfaces—interfaces have agreed magnitude and units of measure
- Scope of change clearly identified.

Robustness

Robustness is an approach/methodology that allows the engineer to analyze functions performed by the system in order to appropriately define characteristics that control behavior and identify uncontrolled influences—noise. Describe transformation: make sure the physics of the design product is understood (P-Diagram is usually the tool of choice to identify the inputs, transformation, output, errors, controls, and noises as well as learn influences on performance). Furthermore, with robustness one is able to assess testing (robustness checklist) in order to confirm testing represents "real" situations and ensure testing that includes the right characteristics. In essence, robustness elements describe how the transformation will be delivered, through a definition of characteristics to control in design, manufacturing, and testing. The best transformation of course is the *ideal function* that transforms *all* inputs to outputs. In this case there are no errors, controls, or noises to contend with.

P-diagrams

Why

- Focus on achievement of function.
- Clearly identify all influences on that function—what can be controlled (control factors) and what can't reasonably be controlled (noise factors).
- Key input for robust DFMEA analysis.

Mindset

- Focus on engineering, i.e., design decisions necessary to optimize output; relationships between inputs and outputs, and link noises and specific output changes—both magnitude and performance

- Selective application—select specific functions to analyze robustness failures or new technology
- Preparation for DFMEA—assess understanding of transformations and how they change and identify countermeasures and influences required in testing

Best Practice

- *Function specific*: use one diagram per function and only key functions to be analyzed
- *Balanced transformation*: conservation of material, energy or information can be seen
- *Failures and losses are partitioned:* make sure failures of function are measured in units of desired output and are indeed measurable—diversion or losses are described as separate outputs
- *Noises are measurable and traceable*: noises are capable of description in physical units, and sources of noise will be seen in the interface analysis and boundary diagram

Target response

Design parameters vary to optimize system performance. That is: control factors or parameters are tuned to perform *function* and to a target *response*. Remember that *all* performance characteristics need to be considered.

Robustness checklist (RCL)

Why

- Understand interactions between noise factors and failure modes
- Identify and improve testing to ensure coverage of noise–failure mode interactions

Mindset

- For a few important functions, the RCL is a *systematic* means of assessing the usefulness of test and assessment methods
- Keep the assessment simple by considering only those functions that have experienced robustness issues previously or are at risk due to changes in the application or operating environment
- RCL can be used to justify ratings of detection controls in DFMEA

Best Practice

- RCL will focus on a few functions and failure modes where justified by quality history or risk assessment

- Simple RCLs are easier to understand and allow teams to concentrate on analysis, not managing the tool
- Values for noises applied in testing should be confirmed as being appropriate to the program application
- Opportunities for test improvement, e.g., a move from a bogey test to one of continuous measurement of function versus noise should be identified
- Test improvements that are found should be included as recommended actions in the DFMEA and confirmed in design reviews

Failure analysis (FA) and mistake avoidance (MA)

Typical issues here are to: (a) *describe failures* that may and have occurred, (b) *assess consequences* and reasons for failure, and (c) explain why *design or testing* won't allow failures to escape. Common approaches to these issues are to:

- List functions—describe performance; include "unspoken" requirements
- Describe failures—failure to perform—magnitude; situation and condition—not "category"
- Prioritize issues—high severities > high occurrence
- Detail causes—specific mechanisms for failure modes
- Confirm countermeasures

DFMEA

Why

- Manage risk
- Evaluate design controls
- Identify actions to reduce risk
- Document the analysis

Mindset

- Thorough analysis means that *all* functions are considered; specific engineering terms are used to define all items, functions, causes and controls. Furthermore, the terms used to identify the unknown, not just document existing knowledge.
- Controls: both detection and prevention controls identified and gauged to useful life.
- Actions: all actions must take into consideration, so prioritization is always based on risk and the ultimate action will reduce severity, occurrence, or detection, i.e., improve the design or improve the testing.

DFMEA Backbone of FMA

- DFMEA ties all the other failure mode avoidance tools together.
- Documents the key outcomes in a single analysis.
- Some tools are inputs to the DFMEA.
- Some tools are supporting tools during DFMEA execution.
- Some tools are outputs of the DFMEA.
- Some tools can be used as both interchangeably.

Communication

As we already have discussed, communication at all levels is imperative for success. For example, in the area of special characteristics, open discussion is required for the appropriate care in production. Furthermore, the appropriate and applicable interactions (interfaces) with other systems must be understood to: (a) identify important characteristics (i.e., (i) what product characteristics prevent failure? (ii) which are influenced by manufacturing? (iii) which manufacturing facility best controls them?) and (b) agree on the specific controls for those characteristics (i.e., (i) controls will be specific to manufacturing facility and (ii) often controls are already in place). In both cases, identifying specific characteristics allow manufacturing to control efficiency (i.e., credible suppliers will have controls to prevent shipment of non-compliant product).

Verification

Demonstrate performance that is always—as required—verified through at least two approaches:

1. Analytical and physical testing: (a) verify performance and (b) verify absence of failure modes.
2. DVP is reviewed through FMA tools: (a) ensure all failure modes and causes are covered, (b) ensure noises correctly represent reality, (c) initial DVP feeds into FMA tools, and (d) DVP is refined and improved through FMA tools.

Note that QH and robustness elements identify opportunities to improve testing. Remember that often simple adaptions of existing methods can be done at minimal cost.

DVP&R including PV

Why

1. Verify the design meets requirements in the presence of all significant noise factors.

2. Verify production parts meet requirements in the presence of all significant noise factors.

Design Verification (DV): Analysis or testing executed to demonstrate that new or modified designs, or new applications of existing designs, will meet customer expectations, in the intended environment, over the useful life of the product.

Production Validation (PV): Analysis or testing executed to validate that products made from production tools and processes will meet customer expectations, in the intended environment, over the useful life of the product.

Mindset/Best Practice

- Requirements—every requirement verified
- Two-phase verification: (a) verification prior to design release and (b) verification prior to design completion
- Noise factors included, including useful life
- Sample size—statistically meaningful
- Acceptance criteria—measurable engineering terms
- Critical component tests identified and timed to complete before M1/VP MRD
- PV—full PV program defined through ES and eFDVS

When to Apply: (a) Required for all systems/components based on a change point approach, and (b) analysis focused on what has changed

Format: The format should be identified in the organization's procedures and instructions, and it should be followed in the internal documentation.

Responsible: The responsible person should be the D&R engineer.

Support: All successful programs have appropriate and applicable support. DVP&R is no different. Typical support is from: function quality, attribute leads, technical specialist, and suppliers.

References

Advisera. (2017). https://advisera.com/16949academy/blog/2017/09/13/establishing-advanced-product-quality-planning-apqp-in-iatf-16949/. Retrieved on March 13, 2018.

AIAG. (2005). *Statistical Process Control: SPC.* 2nd ed. Southfield, MI: Chrysler, LLC, Ford Motor Co. General Motors.

AIAG. (2008a). *Advanced Product Quality Planning and Control Plan: APQP.* 2nd ed. Southfield, MI: Chrysler, LLC, Ford Motor Co. General Motors.

AIAG. (2008b). *Potential Failure Mode and Effects Analysis: FMEA.* 4th ed. Southfield, MI: Chrysler, LLC, Ford Motor Co. General Motors.

AIAG. (2009). *Production Part Approval process: PPAP.* 4th ed. Southfield, MI: Chrysler, LLC, Ford Motor Co. General Motors.

AIAG. (2010). *Measurement Systems Analysis: MSA.* 4th ed. Southfield, MI: Chrysler, LLC, Ford Motor Co. General Motors.

AIAG. (September 2017). *MAQMSR (Automotive Quality Management System Document).* 2nd ed. AIAG, ANFIA, FIEV, SMMT, and VDA – 2016.

American Society of Mechanical Engineers (ASME). (2009). *ASME Y14.5:2009.* ASME. NY.

Automotive Industry Action Group/Verband der Automobilindustrie (AIAG/VDA). (2017). *Failure Mode and Effect Analysis: FMEA (Design FMEA and Process FMEA Handbook).* 1st ed.

Bales, R. F. (1965). The equilibrium problem in small groups. In A. P. Hare, E. F. Borgatta and R. F. Bales (Eds.) *Small Groups: Studies in Social Interaction.* New York: Knopf.

Breyfogle, F. (2003). *Implementing Six Sigma: Smarter Solutions Using Statistical Methods.* 2nd ed. New York: John Wiley & Sons.

Brown, R. (1999). *Group Processes.* 2nd ed. Oxford: Blackwell.

Carr, M., S. Konda, I. Monarch, C. F. Walker, and F. Carol Ulrich. (1993). Taxonomy-based risk identification. Software Engineering Institute. *CMU/SEI Report Number:CMU/SEI-93-TR-006.* https://resources.sei.cmu.edu/library/asset-view.cfm?assetID=11847. Retrieved on March 30, 2018.

Committee Draft of ISO 31000 Risk management (PDF). *International Organization for Standardization. 2007-06-15.* Archived from the original (PDF) on 2009-03-25. Retrieved on April 23, 2018.

Common Vulnerability and Exposures list. Cve.mitre.org. Retrieved on April 23, 2018.

Corrective Action Preventive Action (CAPA). (2018). https://quality-one.com/capa/. Retrieved on May 15, 2018.

Courtney, R. Jr. (IBM, 1970). https://www.coursehero.com/file/p11dts6/In-business-it-is-imperative-to-be-able-to-present-the-findings-of-risk/. Retrieved on April 30, 2018.

Crockford, N. (1986). *An Introduction to Risk Management*. 2nd ed. Cambridge, UK: Woodhead-Faulkner. p. 18.

Dorfman, M. (2007). *Introduction to Risk Management and Insurance*. 9th ed. Englewood Cliffs, NJ: Prentice Hall.

Durivage, M. (March 2017). Not harder: Make your CAPA verification of effectiveness SMART. *Quality Progress*. 41–43.

FCA US LLC. (October 17, 2016a). *Customer-Specific Requirements for PPAP*. 4th edition and Service PPAP. 1st Edition.

FCA US LLC. (October 17, 2016b). Customer-Specific Requirements for IATF 16949:2016.

Ford Motor Co. (2011). *FMEA Handbook*. 4.2 ed. Dearborn, MI: Ford Motor Company.

Ford Motor Co. (June 2013). Ford Motor Company Customer-Specific Requirements for use with PPAP 4.0

Ford Motor Co. (March 22, 2017). Launch of Ford Customer Specifics to IATF 16949.

Ford Motor Co. (May 1, 2017). Ford Motor Company Customer-Specific Requirements for IATF-16949:2016.

Forsyth, D. R. (1990, 1998). *Group Dynamics*. Pacific Grove CA: Brooks/Cole Publishing.

General Motors. (2016). *IATF 16949: Customer Specific Requirements*. General Motors Company.

General Motors. (November 28, 2017). IATF Guidance on the application of Customer Specific Requirements (CSR) and Supplier Codes.

General Motors. (December 2017). IATF Oversight Certification Body Communiqué: CB COMMUNIQUE # 2017-012.

Heathfield, S. (January 14, 2018). What are the stages of team development. https://www.thebalancecareers.com/what-are-the-stages-of-team-development-1919224. Retrieved on May 9, 2018.

https://www.smartsheet.com/8-elements-effective-change-management-process. Retrieved on May 18, 2018.

IADC HSE Case Guidelines for Mobile Offshore Drilling Units 3.2, section 4.7.

Jacob, D. (July 24, 2017). Quality 4.0: Fresh thinking for quality in the digital era. *Quality Digest*. https://www.qualitydigest.com/inside/management-article/quality-40-072417.html. Retrieved on May 21, 2018.

Jocz, W. (May 14, 2018). Discussion with Chief Engineer - Global FMA. Ford Motor Company.

Kokcharov, I. (June 2, 2015). What is project management? https://www.slideshare.net/igorkokcharov/what-is-project-risk-management. Retrieved on May 18, 2018.

The Lean Enterprise. http://www.freeleansite.com/. Retrieved on May 15, 2018.

Llopis, G. (October 1, 2012). The 12 crucial leadership traits of a growth mindset. https://www.forbes.com/sites/glennllopis/#47d908ff50d0. Retrieved on May 9, 2018.

Lozier, T. (March 21, 2018). *Risk-Based Thinking and the Digital Transformation of Quality*. https://www.qualitydigest.com/inside/risk-management-article/risk-based-thinking-and-digital-transformation-quality-032118.html? Retrieved on March 21, 2018.

McGivern, G. and M. Fischer. (February 1, 2012). Reactivity and reactions to regulatory transparency in medicine, psychotherapy and counseling. *Social Science and Medicine* 74 (3): 289–296.

PMBOK (Project management body of knowledge (Guide)). (2017, 2000). Project Management Institute. Newton, PA.

Roehrig, P. (2006). Bet on governance to manage outsourcing risk. *Business Trends Quarterly.* http://www.objectivasoftware.com/docs/BusinessTrendsQuarterly_Jan_2007.pdf. Retrieved on January 7, 2018.

Simon, P. and D. Hillson. (2012). *Practical Risk Management: The ATOM Methodology.* Vienna, VA: Management Concepts.

Stamatis, D. (1997). *The Nuts and Bolts of Reengineering.* Red Bluff, CA: Paton Press.

Stamatis, D. (1998). *Advanced Quality Planning: A Commonsense Guide to AQP and APQP.* New York: Quality Resources.

Stamatis, D. (2002). *Six Sigma and Beyond: SPC.* V.4. Boca Raton, FL: CRC Press.

Stamatis, D. (2003). *Failure Mode and Effect Analysis: FMEA from Theory to Execution.* 2nd ed. Rev. and expanded. Milwaukee, WI: Quality Press.

Stamatis, D. (2002–2003). *Six Sigma and Beyond.* Vol. 1–7. Boca Raton, FL.

Stamatis, D. (2010). *The OEE Primer: Understanding Overall Equipment Effectiveness, Reliability, and Maintainability.* Boca Raton, FL: CRC Press.

Stamatis, D. (2014). *Introduction to Risk and Failures: Tools and Methodologies.* Boca Raton, FL: CRC Press.

Stamatis, D. (2015). *The ASQ Pocket Guide to Failure Mode and Effect Analysis (FMEA).* Milwaukee, WI: Quality Press.

Stamatis, D. (2016). *Quality Assurance: Applying Methodologies for Launching New Products, Services, and Customer Satisfaction.* Boca Raton, FL: CRC Press.

Tuckman, B. W. (1965). Developmental sequence in small groups. *Psychological Bulletin* 63: 384–399.

Tuckman, B. W. and M. A. C. Jensen. (1977). Stages of small group development revisited. *Group and Organizational Studies* 2: 419–427.

Wheeler, D. (2010). *Understanding Statistical Process Control.* Knoxville, TN: SPC Press.

ISO Standards

ISO: International Organization for Standardization ISO Central Secretariat. Geneva, Switzerland. (ISO is NOT an abbreviation of the "International Organization for Standardization." If it was a) it would be IOS and b) it would have periods after each letter. ISO is a Greek word which means "equal" and implies inherently a "harmony." The International Organization for Standardization chose it to communicate the idea of a "family of standards" that it will be uniquely recognized worldwide as a single "quality family" of standards. To that end, it has been very successful).

AS9100D: Quality Management Systems—Requirements for Aviation, Space, and Defense Organizations.

IATF 16949:2016—Quality management system for organizations in the automotive industry.

ISO 13485:2016—Management Software Systems.

ISO 14001—Environmental Management Systems.

ISO 14001:2015—Environmental Management Systems.
ISO 14971—Automated Risk Analysis & Management Software.
ISO 16792 3D (2015). *Technical Product Documentation: Digital Product Definition Data Practices*. 2nd ed. Geneva, Switzerland: ISO.
ISO 45001:2018. *Occupational Health and Safety Management Systems: Requirements with Guidance for Use*. Singapore: Enterprise Singapore.
ISO 9001- specifies requirements for a quality management system.
ISO 9001:2015—Quality management system.
ISO/DIS 31000. (2009). *Risk Management: Principles and Guidelines on Implementation*. International Organization for Standardization.
ISO/DIS 31000. (2018). *Risk Management: Principles and Guidelines on Implementation*. International Organization for Standardization.
ISO/IEC 17025. (2017). *General Requirements for the Competence of Testing and Calibration Laboratories*.
ISO/IEC Guide 73:2009. (2009). *Risk Management: Vocabulary*. International Organization for Standardization.
ISO/TS 16949. (2009). *Quality Management Systems. Technical Specification*. Southfield, MI: ANFIA, FIEV, SMMT, VDA, Chrysler, LLC, Ford Motor Co. General Motors, PSA Peugeot Citroen.

Risk Management Standards

There are three main Australia and New Zealand standards that cover Risk Management:

AS4360:2004. *Risk Management*. New Zealand: Standards Australia/Standards.
HB221:2004. *Business Continuity Management*. New Zealand: Standards Australia/Standards.
HB436:2004. *Risk Management Guidelines: Companion to AS/NZS 4360:2004*. New Zealand: Standards Australia/Standards.

Selected Bibliography

Antunes, R. and V. Gonzalez. (March 3, 2015). A production model for construction: A theoretical framework. *Buildings* 5 (1): 209–228.
Brożyńska, M., K. Kowal, L. Anna, and M. Szymczak. (2016). *5-Whys. Method First Handbook*. Łódź, Poland: 2K Consulting. p. 34.
Chrysler Corporation. https://www.yumpu.com/user/cda.extra.chrysler.com. Retrieved on October 21, 2018.
DumpsBook Publishers. (2017). *Certified in Risk and Information Systems Control (CRISC): Real Exam Questions*. DumpsBook Publishers. Retrieved on February 23, 2018.
Fantin, I. (2014). *Applied Problem Solving. Method, Applications, Root Causes, Countermeasures, Poka-Yoke and A3. How to make things happen to solve problems*. Milan, Italy: Createspace, an Amazon company.
"Five Whys Technique." adb.org. Asian Development Bank. February 2009. Retrieved on May 10, 2018.
Flyvbjerg, B. and A. Budzier. (2011). Why your IT project may be riskier than you think. *Harvard Business Review* 89 (9): 601–603.

Garvey, P. R. (2008). *Analytical Methods for Risk Management: A Systems Engineering Perspective*. Boca Raton, London, New York: Chapman-Hall/CRC-Press, Taylor & Francis Group (UK).

Hubbard, D. (2009). *The Failure of Risk Management: Why It's Broken and How to Fix It*. New York: John Wiley & Sons. p. 46.

Husted, E. (March 2018). Model based definition put GD&T data to work. *Quality*. 25–29.

Kossiakoff, A. and W. N. Sweet. (2003). *Systems Engineering Principles and Practice*. New York: John Wiley & Sons, pp. 98–106.

Kotter, J. (1996). *Leading Change*. Boston, MA: Harvard Business School Press. https://www.smartsheet.com/8-elements-effective-change-managementprocess.

Ohno, T., foreword by Norman Bodek. (1988). *Toyota Production System: Beyond Large-Scale Production*. Portland, OR: Productivity Press.

Rodriguez-Perez, J. (2016). *Handbook of Investigation and Effective CAPA Systems*. Milwaukee, WI: ASQ Quality Press.

SAE. (2009). *J1739: Potential Failure Mode and Effects Analysis in Design (Design FMEA), Potential Failure Mode and Effects Analysis in Manufacturing and Assembly Processes (Process FMEA)*. Warrendale, PA: Society of Automotive Engineers (SAE).

Thurer, M., A. Stevenson, and C. Protzman. (2016). *Card-based Control Systems for a Lean Work Design: The Fundamentals of Kanban, ConWIP, PPOLKA and COBACABANA*. New York: Productivity Press.

Wilson, B. (September 24, 2014). Five-by-Five Whys. www.bill-wilson.net. Retrieved on May 10, 2018.

Printed in the United States
by Baker & Taylor Publisher Services